THE PARROT IN THE MIRROR

THE PARROT IN THE MIRROR

how evolving to be like
birds made us human

ANTONE
MARTINHO-TRUSWELL

OXFORD
UNIVERSITY PRESS

Great Clarendon Street, Oxford, OX2 6DP,
United Kingdom

Oxford University Press is a department of the University of Oxford.
It furthers the University's objective of excellence in research, scholarship,
and education by publishing worldwide. Oxford is a registered trade mark of
Oxford University Press in the UK and in certain other countries

Impression: 2

Published in the United States of America by Oxford University Press
198 Madison Avenue, New York, NY 10016, United States of America

British Library Cataloguing in Publication Data
Data available

Library of Congress Control Number: 2021942678

ISBN 978–0–19–884610–9

DOI: 10.1093/oso/9780198846109.001.0001

Printed and bound by
CPI Group (UK) Ltd, Croydon, CR0 4YY

+AMDG+

To Emma, Flora, and Clara,

For whom I do all things

And

To my parents, and to Alex K:

It is as much their work as mine.

CONTENTS

INTRODUCTION

Sixty-six million years ago, an asteroid crashed to Earth in what is now Mexico, and brought the age of reptiles to an end. Massive reptiles, including the dinosaurs on land, as well as similar flying and swimming species, had dominated the Earth for a long time—about 150 million years. At a stroke, the dinosaurs were doomed to oblivion. Worse than that! It is now thought that every tetrapod—that is, four-limbed animal—weighing over 25 kilogrammes died in the years following the impact, with a few rare exceptions like the leatherback turtle.

In the ashes of a world undone, a new set of animals came to the fore. Previously an undignified group, a sideshow to the reptiles' dominance, the newly scoured Earth gave them the opportunity to explode in diversity and capability, evolving new forms and developing new capabilities. They were carnivores and herbivores, hunters and hunted. Some swam, some grew larger, some became ferocious.

On one continent, where conditions were just right, they became very smart. Their lives grew longer, and their social interactions more complex. They stood upright, and talked a great deal more than their close relatives. Their numbers were small as yet, but over time, equipped with big, creative brains and seeking out resources, they began to roam. Eventually, they found the bridge off their continent and spilled out into the rest of the world. They spread quickly, over all the continents. Soon, the smartest animals anywhere on Earth had populated most of the planet.

And fifty million years after the parrots had accomplished all this, humans arrived, and did it again.

* * *

Parrots are probably humans' favourite birds, and it is easy to see why. They are colorful and playful, fascinating to watch, have vaguely human facial expressions and body language, and, of course, they talk. That would be enough for us to be fascinated by them. But I think there is something more. Parrots are our evolutionary mirror image. We are only very distantly related, in the sense that all vertebrates are, but share many similarities—more so, in some cases, than with our closer animal relatives. Their evolutionary history follows similar patterns, and they are nature's other attempt at extraordinary intelligence, in a distant branch of the tree of life. These similarities can be surprising, given how fundamentally different we seem to be from birds. I would argue, though, that even beyond the parrots, we humans can find great similarities with many birds that we do not share with other mammals. Indeed, we are positively bird-like.

Humans are animals, more specifically, mammals, and, more specifically, primates. Primates are a group of highly intelligent, versatile mammals that originated in Africa and include the great apes—ourselves, chimpanzees, bonobos, gorillas, and orangutans—as well as all the other monkeys, baboons, and similar animals. The primates all have thumbs, and our hair and internal organs are more or less like those of other mammals. We are not closely related to birds, and we do not look anything like them. Our bodies, our eyes, our skin and hair, our blood, our brain structure, and our ancestry, diverge radically from the birds.

Yet, in another way, we humans also have many things in common with birds. The ways of living first 'invented' by birds are very different from typical mammal strategies, but very similar to those that later made primates unique among mammals, and humans unique among primates.

I suspect this is part of why we find birds so captivating. One might think that when an animal has become intelligent enough to start studying itself and other animals, its focus would be on its nearest relatives. To some extent, that is true. Most of the animals that we live closely with are mammals that we have domesticated, like dogs, cats, and livestock (other than poultry). A great deal of scientific research is focused on rats and other rodents (to better understand the mammal body, in a cheap and expendable form), or our close relatives, the chimpanzees and other primates (to better understand the mammal brain, in an expensive and precious form).

Yet there is an enormous interest in birds, especially among animal lovers and amateurs. Aviaries, birdwatching, nest-boxes, birdfeeders, pets, clubs—we love birds. Birdsong in the morning is such a peaceful and satisfying background that we have devised alarm clocks that can wake us with a simulated dawn chorus. Birds are our symbols: we have doves for peace and eagles for patriotism. People describe themselves as bird lovers, and birdwatchers keep 'life-lists'—with each new bird species sighted becoming a lifelong accomplishment to be cherished and compared with other enthusiasts. Grandparents take children to feed ducks in the park, fanciers keep pigeons for racing and homing. And I suspect every human, alive or dead, has had that dream—you know, the one where you can fly, like a bird.

Biologists also love birds. Our modern understanding of animal behaviour was established partly thanks to Konrad Lorenz's work on geese and ducks. Lorenz arranged for groups of baby ducks and geese to see him as the first moving object in their lives instead of their and he did a number of now-famous experiments like quacking to them to encourage them to follow him. His discovery of imprinting—the process by which a newborn chick or duckling learns to recognize its mother—would win him a Nobel Prize for Physiology or Medicine in 1973. Nowadays, newly hatched chickens are a standard model organism for studying all kinds of behaviour and physiology in the lab, as well as in the biology classroom. Pigeons have a long history in experiments on the process of learning, and in the study of animal navigation. B. F. Skinner used pigeons in his famous 'Skinner box' experiments, where he showed that you could train an animal through precise rewards to carry out almost any action. He even managed to teach two pigeons to play a sort of ping-pong/air hockey hybrid game by rewarding the scores with birdseed. Studies on great tits and other small songbirds helped to establish the field of ecology and now underpin much of what we know about behavioural economics and how animals optimize their behaviour. Add to this a variety of famous work on intelligence in parrots and crows, and huge amounts of conservation-focused study, and it is clear that the biological profession has a healthy interest in birds.

There are some straightforward reasons for all this interest. Most birds are diurnal—that is, they are awake during the day, like us—while most mammals come out at night, meaning most people see many more birds in their gardens or local parks than they do mammals. This also makes birds convenient for scientists. If the question you want to answer can be tested with any animal, it is convenient to choose one that means you do not have to be in the lab at midnight, or keep the place dark at noon. In some cases, as when studying baby chickens, obtaining the animals is also convenient: chicken eggs hatch to provide the world's most readily available animals, thanks to our extreme love of eating them. And unlike baby rats, which must be birthed by their mother and are difficult to hand-rear, chicks can be artificially incubated and hatched, and can walk, eat, and drink on their own within hours of hatching.

It is also easy to attribute our love of birds simply to their charisma. Most birds are beautiful, or at least interesting to look at and watch. Many, especially the larger species, such as the ducks you see in parks, or city pigeons, are not terribly afraid of us, and quite willing to go about their business in our presence—making them more entertaining than a rat or deer that flees when humans approach. In general, there are also more obviously different species of wild or feral birds that can be glimpsed by city and suburb dwellers, so there is more interest to be had in comparing one bird to another than in tallying up yet another rat or cat. And many birds have impressive tricks with which to delight us, from parrots' mimicking, to peacocks' tail displays, to geese flying in formation. These behaviours and traits certainly account for some of our species' collective bird obsession.

Yes, it could just be that we like birds because we are curious and birds are inherently interesting. I am an ornithologist, after all, so I might be expected to find them *very* interesting. But I think the affinity runs deeper.

I think we see ourselves in birds.

I don't just mean this in the obvious ways—the anthropomorphism that comes from looking at a parrot, standing upright, with big, expressive eyes, and bobbing its head around to get a better look at you. More than this, there are deep similarities in the way our evolutionary histories have played out, and our different bodies and brains have hit on similar strategies and traits, despite all our differences. So much so, that in many meaningful ways, especially our behaviour, we humans have given up living like mammals in favour of living more like birds.

By discovering, adopting, and adapting the strategies that birds had discovered before us, we became the bird without feathers.

* * *

When we want to understand the plumbing and infrastructure of the human body, we have to look at mammals—birds are so physically different that it can be hard to draw conclusions, while a rat's internal workings, or better yet, a pig's, can give us a really good idea of our own. But when we set out to understand our minds, our behaviour, our social lives, or the reasons why our evolutionary history has unfolded in a particular way, it is to the birds, and especially the songbirds and parrots, that we should look. We are truly exceptional mammals. The similar bodies of primates and rats belie the fact that these animals live radically different lives, and humans got where we are by avoiding some very common mammalian behaviours.

Compared to birds, though, we are not so exceptional. Where mammals zig, and birds zag, we can be depended upon to zag with the birds. Our lifespans, our breeding, our minds, our relationships, and our history all fit neatly the bird exemplar—sometimes becoming the best example of a bird-ish trait. By studying birds, we therefore better understand the evolutionary forces that shaped humans, and thus better understand ourselves. It can be easy to dismiss research on the intelligence of crows or the social behaviours of cowbirds as interesting but not truly useful—enter taining sprinkles on the hard, practical cake of biomedical science. This misunderstands one of the most important parts of science.

If we want to really get to grips with something in the natural world, then we need to answer two questions: 'how' and 'why'. Studying and understanding mammals is the only way to answer the 'how' questions about humans: how does our heart work?, how do we grow and develop?, how do we fix broken bodies? Birds are too different from us to answer these questions. Studying birds, though, can give us insight into the 'why' questions—sometimes even insights that mammals might not be able to offer. This is why we should, at least sometimes, treat humans as the birds without feathers. As the following chapters will demonstrate, this can give us some remarkable insights into our first and favourite question about ourselves: why?

1

HOW DID WE GET HERE?

When Charles Darwin first wrote about the theory of evolution in *The Origin of Species*,[1] he drew his argument to a close with an eloquent description of a 'tangled bank'. Darwin thought of a riverbank, filled with plants, insects, birds, mammals, and countless other species, all so very different from each other, having arisen from a single ancestor, and marvelled at how such diversity of life could be produced by the simple processes of evolution that he had described. This diversity is still what draws biologists and enthusiasts (indeed, perhaps all of us) to the natural world.

Before Darwin had even started on his work—the work that essentially founded the modern field of biology—another early biologist hit on a key component of Darwin's theory. Carl Linnaeus was a Swedish biologist who became the very first taxonomist. Taxonomy is the study of biological groupings and classifications. The vast diversity of life that Darwin was later to write about needed to be categorized and classified if it was going to be rigorously studied, and Linnaeus built the first version of our modern system of taxonomy, which he published in 1735 in his book *Systema Naturae*.[2] Linnaeus created a series of groupings of decreasing size and specificity into which he sorted all the life forms known at the time, including some, like sea monsters, that we now know to be myths—or, at least, misinterpretations of the truly enormous real animals like whales and giant squids that do lurk in our oceans. He started with five groupings, though we now have many more, and he recognized that these groups were constructs, i.e., where we grouped an animal or plant was not incontrovertible truth, but our best attempt at placing it with meaningfully similar organisms. Most importantly, his groups were nested, so that an individual *species* was grouped into a *genus* with several very similar species. Each genus, and all the species within it, were in turn grouped into an *order*, each order into a *class*, and each class into a *kingdom*. At the time, Linnaeus was building this hierarchical system based mostly on physical similarities—he did not yet have Darwin's theory to reveal the connectedness of all of life. But inadvertently, he was, in most cases, grouping organisms by how closely related they were—all before anyone even knew that different species could be related to each other.

Linnaeus may likely have thought of his groupings in terms of concentric, nested circles, giving us increasingly specific information about each smaller subdivision, but you could just as easily map them out as a tree (Figure 1). That tree formation is what Darwin ultimately hit upon. He recognized that all of life could be fitted into a tree diagram; and not just a descriptive tree, but a family tree.

This meant that Linnaeus's two living kingdoms, animals and plants (he had originally included a third, minerals) were themselves branches of the one related family of all life on Earth. Scientists would later add several other kingdoms describing fungi and micro-organisms, and more levels, like the *phylum* and the *family*, to increase the precision of groups. They recognized that all

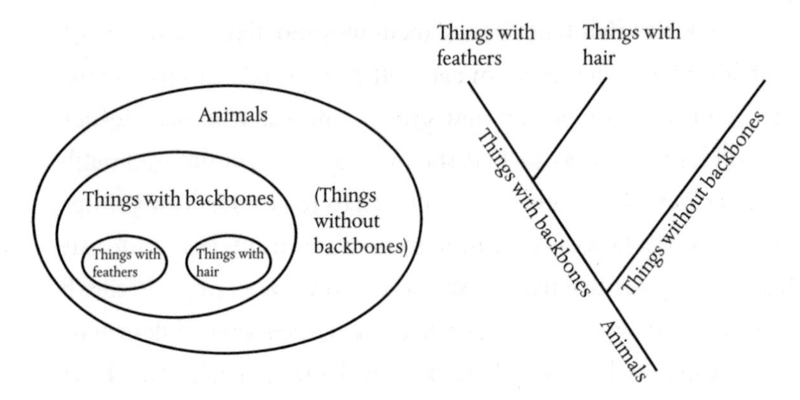

Fig. 1 Linnaeus's categories were nested, much like in the diagram on the left. His multiple layers of groupings could describe an organism that was an animal, with a backbone, and feathers, and group it more closely with other animals with a backbone and hair than those with no backbone at all. The same nested groupings can also be modelled using a tree (right).

these kingdoms are branches diverging ultimately from a single trunk, from a shared original common ancestor from which all organisms evolved and diversified.

Linnaeus's system allowed the relationships between species that Darwin discovered to be structured and described, and it is still used today, constantly modified and argued over as scientists gather better data about genetics, morphology, and history to reclassify and better describe organisms. Linnaeus's initial reliance on how animals looked and behaved has given way to molecular taxonomy, which allows us to directly compare the DNA of living species to determine which ones are more closely related. For example, Linnaeus had grouped together all of the invertebrates except the arthropods (insects, spiders, and crustaceans), into a group he called 'Vermes'. Linnaeus described these as 'animals of slow motion, soft substance . . . the inhabitants of moist places'.[3]

We now know that his 'Vermes' includes more than 30 groups of animals all as different from each other as insects are from reptiles, and we have created new groups and subgroups to reflect that diversity–groups such as the molluscs, flatworms, and jellyfish. Further DNA studies have allowed us to get closer and closer to the truth of how species, even those superficially very different from each other, are related. We now know, for example, that we primates are part of the magnificently named superorder Euarchontoglires, which we share with rodents and rabbits. Those animals are, perhaps surprisingly, more closely related to us than those in our sister superorder Laurasiatheria, which includes the ungulates, including the horse and the rhinoceros, and carnivores, such as dogs and cats.

The modern tree of the animal kingdom (Figure 2) records the history of how animals grew from the first multicellular organisms into all the groups we know today. As you move up the tree toward the tips, you move forward in time. The very ends at the top represent the present day, and the lines of the branches represent all the ancestors leading to the organisms alive now. The further down in the tree a branch splits off from another, the more distantly related those groups are. Biologists call this kind of tree a *phylogeny*, and it provides a visual representation of the relatedness of organisms. So, all the vertebrates, which converge to a single ancient ancestor, are more closely related to each other than they are to the nearest invertebrates, the echinoderms, which include starfish. And the starfish are more closely related to the vertebrates than they are to insects, which are arthropods—you have to go further down the tree to find the last common ancestor of the two groups: the place where the lines split.

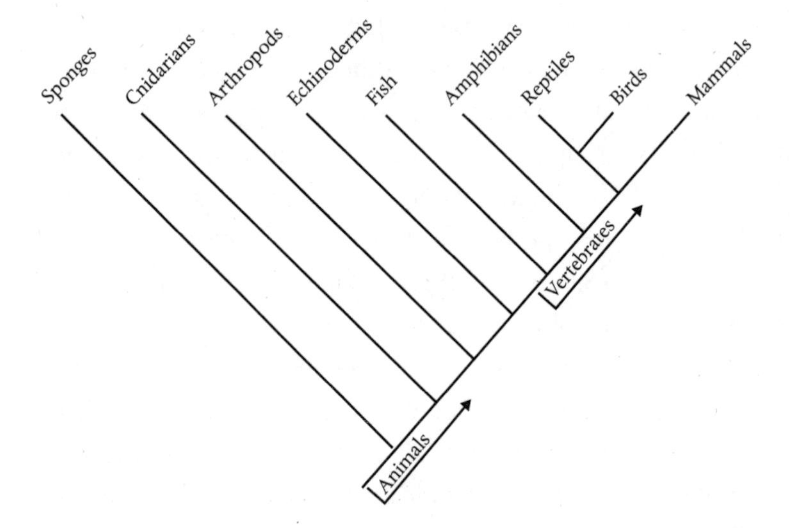

Fig. 2 A very incomplete, simplified phylogeny of the animal kingdom. Groups' relatedness is shown by physical distance: the shorter the distance along the black lines between two groups, the more closely related two groups are. All vertebrate groups are equidistant from the echinoderms, our nearest invertebrate relatives (on this tree); and all vertebrates and the echinoderms are more closely related to each other than any of us are to the arthropods. Many groups have been left out, especially between the Cnidarians and the Echinoderms.

This is how we can make meaningful statements about which animals are closely related and which are more distant, and it shows how birds and mammals, including humans, are related to each other. Figure 3 shows just the vertebrates, all the organisms with a rigid backbone. Vertebrates originated from the invertebrates with the evolution of tunicates—tube-like animals that live a mostly sedentary life attached to rocks (Figure 4). They start life as a free-swimming larva with a tiny brain and a stiffened coating around the main nerve that runs down the length of

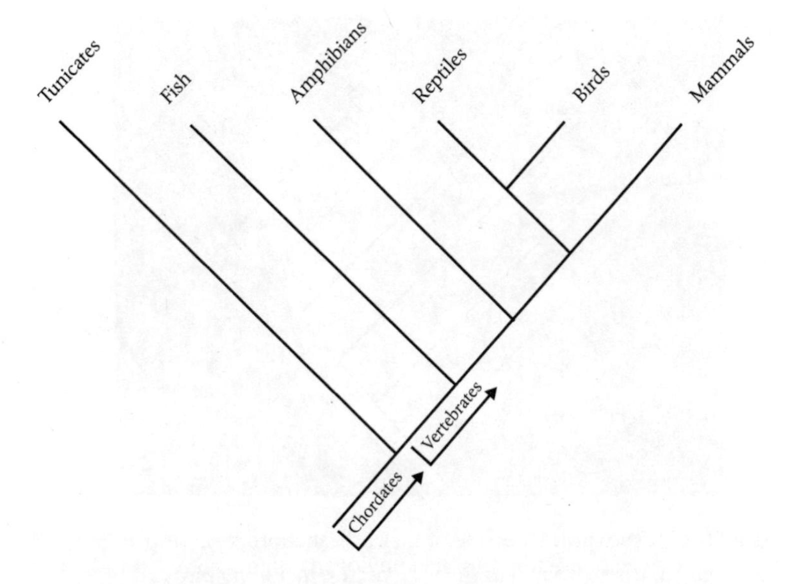

Fig. 3 The vertebrates descend from our common ancestor with the tunicates, which have an early form of backbone called a notochord. All animals with either a true backbone (the vertebrates) and those with a notochord form a group called the 'chordates'.

their body—the early form of a backbone, known as a notochord. When they become adults, they attach themselves head first to a rock, absorb their own brain, and spend the rest of their lives filtering nutrients from the surrounding water and releasing sperm and eggs into the surrounding water to breed. Humble beginnings for the vertebrates.

Eventually, the first true vertebrates diverged from the tunicates, evolving to effectively remain in the free-swimming larval stage, and declining to eat their own brain. These would ultimately become the very early fish. These first fish were a bit like hagfish and lampreys. They did not have jaws or bones and had primitive circulatory systems, much like the invertebrates. Bony

Fig. 4 The sessile tunicate is one of the earliest chordates, the group that would in turn give rise to the vertebrates, including birds and mammals. Despite being more closely related to us than any of the invertebrate phyla, the tunicate is very simple, lacking eyes, head, and other sense organs once it leaves its larval stage.

jaws evolved in the sharks to give their mouths more support for chomping through prey, but they retained cartilaginous skeletons and backbones. Bones proved useful for supporting a resilient body and protecting the increasingly complicated internal organs that fish were evolving. So, the bony fish, with fully hardened skeletons, came next and are now the largest single group of vertebrates. Then, from among the bony fish, the vertebrates finally looked to colonize the land.

Amphibians, such as frogs and salamanders, get their name from the fact that they live part of their lives in the water and part on land. They start off as fish-like larvae—what we know as tadpoles in the case of frogs—and then metamorphose into their adult forms: reabsorbing their swimming tails, sprouting legs, and closing off their gills and replacing them with lungs. This process is quite like watching the evolution from fish to amphibians speeded up within a single animal's lifetime. At first, some fish probably developed the ability to squelch about in moist mud, like modern-day mudskippers, as a way to take advantage of food sources that their water-bound cousins could not access. Over time, having sturdy limbs and the ability to breathe without water proved helpful, and evolution selected for these traits, leading to the amphibians. But amphibians are still tied to water. Their young hatch like fish, and their eggs have no watertight coating: they have to remain immersed or they dry out.

To finish the jump to land, the vertebrates had to develop watertight, sturdy eggs that hatched into land-ready young, and this was the start of the reptiles, from which would finally evolve birds and mammals. It is also where things start to get complicated—and it is largely the birds' fault.

On the surface, birds and mammals seem closer to each other than to reptiles. In general, they both have larger brains than the reptiles, and both are covered in dense growths on the skin—feathers and fur. Most importantly, both are warm-blooded, while every other group mentioned so far, including all living reptiles, is cold-blooded. Being warm-blooded is a big advantage for an animal; it means you can regulate your own body heat—burning the energy from food to warm yourself up, rather than having your temperature determined by the air around you. All of the biological and chemical processes in animals are reliant on temperature. Being able to maintain the right temperature through metabolism, rather than moving to a hotter or colder location, opened up new habitats for birds and mammals. This similarity

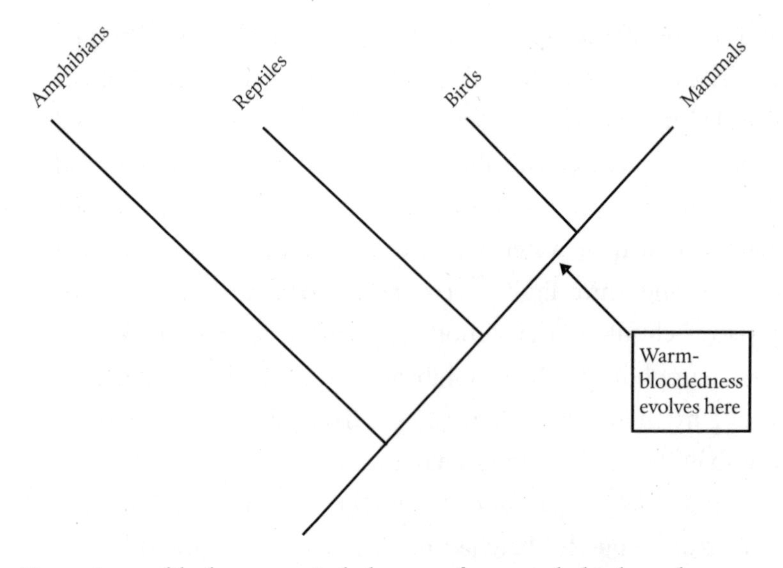

Fig. 5 A possible (but wrong) phylogeny of mammals, birds, and reptiles, based on the assumption that warm-bloodedness evolves only once.

might have suggested a vertebrate tree like Figure 5, with mammals and birds splitting together from reptiles, then splitting from each other later. If you are relying on warm-bloodedness to build the tree, this version is the most likely solution, whereas if birds and reptiles are more closely related than birds and mammals, then warm-bloodedness would have to evolve twice, independently, on separate branches. Evolving any complex trait once is unlikely enough; doing it twice is even harder. For this reason, a shared trait is usually good evidence that it evolved once, in a common ancestor.

But, as we now know, warm-bloodedness *did* evolve twice, and this leads us to one of the most fascinating discoveries about birds. In 1860, a German palaeontologist named Hermann von Meyer found a fossil of a single feather in limestone deposits in Germany. Von Meyer used this single feather to propose a new genus, which he called *Archaeopteryx*—literally 'old wing' or 'old feather' (Figure 6). The following year, a whole skeleton of what appeared

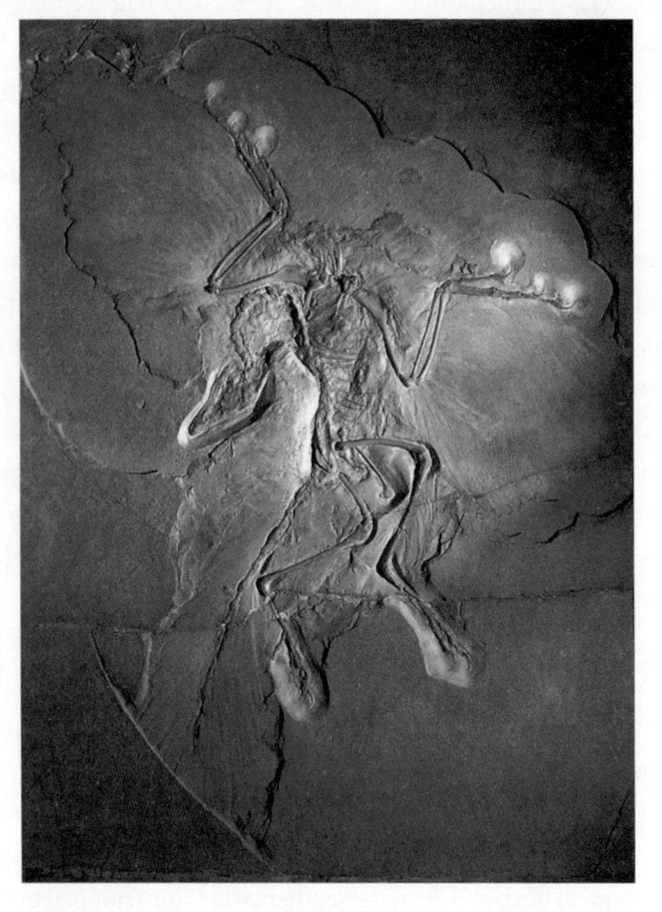

Fig. 6 The fossil *Archaeopteryx* changed our understanding of birds, ultimately revealing them to be dinosaurs that survived the Cretaceous extinction. The fossil *Archaeopteryx* shows the intermediate features of early birds, with a bony tail and clawed proto-wings.

to be a small, early bird species was found, and made its way to the Natural History Museum in London. There, it was classified by the famous English biologist Sir Richard Owen as *Archaeopteryx macrura*—part of the new genus described by von Meyer, and maybe even the same species.

Owen was a difficult man. He was an excellent palaeontologist and skilled at interpreting fossils, and even anticipated some of the details of evolution that would not be confirmed for more than a century. He also coined the word 'dinosaur'. Yet he was a temperamental and occasionally dishonest scientist, claiming credit for others' work, and constantly arguing with Darwin and his supporter, Thomas Henry Huxley, who in turn publicized and emphasized errors in Owen's scientific work. Despite all the bad blood, Darwin cited Owen's *Archaeopteryx* work in new editions of *The Origin of Species*, because the limestone deposits in which the fossils were found were about 150 million years old—much older than had been expected for bird fossils. In fact, the age of the limestone sediments made the *Archaeopteryx* fossils the earliest birds ever discovered at the time—well before the period that Darwin had previously thought birds emerged.[4]

Some scientists were not convinced that *Archaeopteryx* was actually a bird. It was not like modern birds, and still had claws near the tips of its wings, a sturdy, bony tail, and very un-bird-like teeth. This led a few biologists to propose that perhaps feathers, or something like them, had evolved twice, and that *Archaeopteryx* was a reptile that happened to have feathers. It certainly was different from a typical bird, and even the feathers were not arranged quite the same way. The controversy persisted, and it was not until the 1910s that Danish artist-turned-ornithologist Gerhard Heilmann cracked the problem.[5]

Fig. 7 Birds' dinosaur ancestry has been useful to palaeontologists studying how other dinosaurs walked. A scaled-down, weighted tail attached to a chicken serves as a good model for *T. rex*'s gait.

Heilmann compared the *Archaeopteryx* and other birds to a variety of ancient reptiles and came to a stunning conclusion. The group most similar, and from which birds had evolved, were the theropod dinosaurs—a group that includes *Tyrannosaurus rex*. *Archaeopteryx* was a bird, and also, in a sense, a feathered reptile; and feathers had evolved just the once, as a form of modified reptile scales. Heilmann's discovery would be confirmed and confirmed again, and today we have all sorts of evidence for it. Early feathers have been found on dinosaur fossils in China,[6] and we now believe that many different dinosaur species, apart from the bird ancestors, may also have had feathers.[7]

As a result, we now know that birds *are* dinosaurs—the only ones that survived the asteroid impact and subsequent climate change that killed off all the other groups.[8] Every modern bird, from the chicken to the hummingbird to the eagle, evolved from early theropods that developed feathers and started flying. They are all more closely related to the terrifying *Tyrannosaurus* than to any other group of living animals. *Tyrannosaurus rex* may also have been covered in feathers—rather spoiling its popular image. In fact, one of the ways scientists have studied how *T. rex* might have walked is by attaching a prosthetic tail to a chicken (Figure 7).[9]

Birds, then, are surviving dinosaurs, and there is evidence that, unlike modern reptiles, the dinosaurs were warm-blooded. What about the similarities with mammals? It turns out that a different group of reptiles, only distantly related to the dinosaurs, also discovered the benefits of having a protective and warming covering over the skin, and evolved hair by a process similar to birds' feathers, from reptilian scales. Warm-bloodedness did evolve twice: once in the dinosaurs, including the birds, and once in the mammals. The similarity came about because the lifestyles and pressures on both groups led independently to their benefitting from becoming warm-blooded, not because their common ancestor already had them.[10]

Recognizing that birds are a subset of dinosaurs, and therefore a subset of reptiles, makes the tree that we have been building somewhat more complicated, because, as you can see in the revised tree (Figure 8), reptiles are not really a coherent group without birds. An important principle in taxonomy is that you should be able to draw a single line, representing a single common ancestor, and define a group as everything 'downstream' of that line. So, in Figure 8 you can see how amphibians, mammals, and even birds can be defined as everything above a defining line drawn on the tree. Reptiles, though, require two lines—one to split them away from the rest of the vertebrates, and one to stop them from including birds. If we scientists were being really consistent, either 'birds' or 'reptiles' as a high-level grouping for vertebrates would have to go, with either birds becoming reptiles, or, more likely, reptiles splitting into multiple groups.

The tendency of different groups of animals to occasionally hit independently on the same solutions is why the correct family

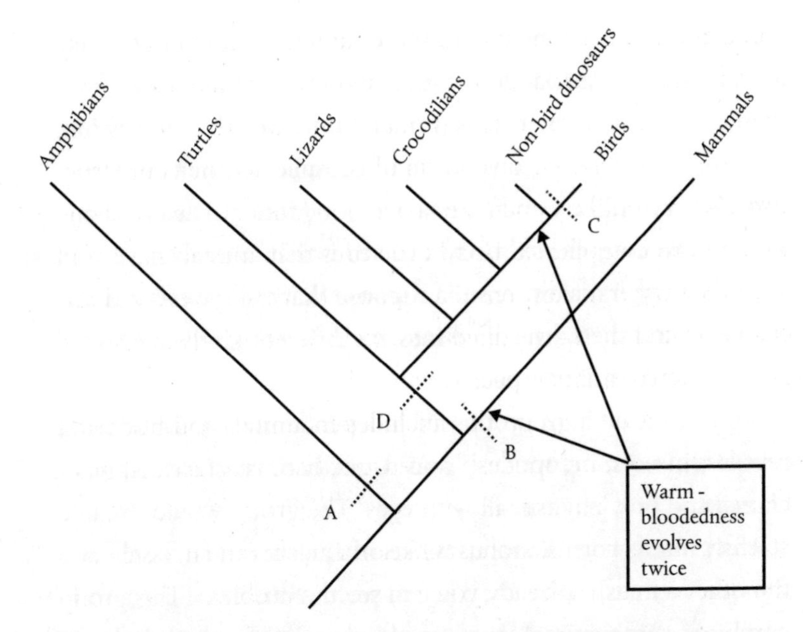

Fig. 8 The correct phylogeny of birds shows that warm-bloodedness did evolve twice, in mammals, and in the dinosaurs. Amphibians (A), mammals (B), and birds (C) can all be defined by a single dashed line, comprising everything 'above' the line in the diagram. Reptiles require two lines, and are all the animals between the dashed line at the base of their branch (D), and the one separating off the birds (C). Note that there is some ongoing uncertainty about the exact positioning of the turtles and lizards—some work suggests they should be swapped.

tree (in Figure 2, and in more detail in Figure 8) is based on much more than just similarity of traits—it is the summation of genetic similarities and our best understanding of when species evolved and when groups split apart. This is important, because, like mammals and dinosaurs, animals that may have very similar anatomy or behaviours can actually be very distantly related. For example, if you were just grouping animals on physical similarities, you might think that a good subgroup would be 'animals with eyes'.

Some animals have them, and some do not, so it could be easy to think that those that do are more closely related to each other than to those that don't. This is not an unreasonable suggestion, because, in general, the evolution of complicated traits or structures is very unlikely and takes a very long time.[11] The eye is one of the most complicated body structures that animals have, and so, at first try, it makes sense to suppose that the eye evolved just the once, and then diversified into the different kinds of eyes we now see across animal species.

So, you create a group that includes mammals and birds and reptiles, insects, octopodes[12] and squid, fish, crustaceans, giant clams, and box jellyfish, all with eyes. The group would exclude starfish, many worms, molluscs like other clams and mussels, and the other jellyfish. Already you can see the problem. This group combines some animals that are closely related—the fish, birds, reptiles, and mammals are all vertebrates—with distantly related others. And it splits closely related groups: giant clams are indeed closer to their smaller, blind cousins than they are to insects, and octopodes and squid are actually molluscs that should be in with the clams, blind and otherwise.

The reason this problem arises is the same as that resulting in the similarity of warm-bloodedness between birds and mammals. This is an effect called *convergent evolution.*[13] While the evolution of complicated traits is very unlikely to happen twice, especially in very distantly related groups, occasionally, a particular trait proves so useful or necessary that it arises independently in more than one group. This is what happened with eyes. The 'camera eye' is the model of eye common to most vertebrates, to cephalopods like octopodes and squids, and even to the box jellyfish. Like a camera, it allows sharp images to be gathered from the surrounding world by focusing light very tightly onto a layer of light-sensitive cells in the back of the eye. Insects have compound eyes, with multiple crystalline receptors that work a bit differently, but still allow them to perceive shapes and colours to get a picture of the surrounding environment. This ability is so useful that despite its complexity, the camera eye has evolved independently, not just twice, but many times, and the eye more generally, in all its different architectures, in as many as forty nine lineages. Many different groups have the same need for accurate vision, to capture prey, avoid predators, and seek a mate, and they have converged on a similar solution for focusing light to form an image. This is convergent evolution.

Convergent evolution is happening all the time, and that is why it is so important that we use both genetics and careful morphological studies to decide how animals are related: if we used superficial similarities, we would constantly be making mistakes.

Being aware of convergent evolution can prevent mistakes, even when you are not using genetics. Think of humpback whales and sea lions (Figures 9a, 9b). Both are mammals that have evolved to live in the sea, and have lost their legs in favour of strong fins that propel them through the water. We know that mammals evolved on land, so ocean-dwelling mammals are a case of land-dwelling mammals re-adapting for swimming. At first, then, you might think that one group of mammals evolved fins and swimming abilities, and then split into the whales and dolphins, on one side (a group called the cetaceans), and the seals and sea lions, on the other (the pinnipeds). But a closer look shows that the fins and swimming are actually the result of convergent evolution, rather than being the ancestral characteristics of both groups. The

Fig. 9 Convergent evolution can lead very different animals to similar-looking solutions. Whales like the humpback are most closely related to hippopotamus and other ungulates, while sea lions derived from the same lineage as dogs and bears. Yet both have lost their land mammal limbs and developed strong fins for swimming. Their superficial anatomical similarities are not a reliable way to build a phylogeny of swimming mammals.

cetaceans have a blowhole for breathing, which is essentially the result of the nostrils having shifted over evolutionary time to be on the top of the head instead of on the front. They have a single rear fin and have entirely lost their leg bones. Meanwhile the pinnipeds have a more conventional mammal nose, with faces looking a bit like dogs, and have hidden remnant leg bones that have evolved to form a tail with, in many species, a split fin. These and other differences reveal the truth: the cetaceans' closest land relative is the hippopotamus,[14] while the pinnipeds evolved from the group Caniformia (literally: 'dog-like'), and are most closely related to either modern bears, or the weasel group, all rather close relatives of dogs (and hence their dog-like faces).[15] Both groups found that moving to the ocean provided them with plentiful food and less competition than on land, so converged on fins and swimming.

This example may give the impression that similarities between distantly related animals would not be very scientifically useful. If we learn about the bone structure and blood vessels in a dolphin's tail, this information will be mostly useless in understanding the structure of a seal's tail, since they evolved separately and with different details. This is indeed a limitation, and it is why, when we use animals in medical research to improve human medicine, we use mammals (usually rats). Animals that are actually closely related tend to have similar physical properties, and we can draw inferences about how one works from the other.

But similarities that come about because of convergent evolution tell us something else. If two very different groups of animals hit on the same solution, despite their very different anatomy and evolutionary background, it tells us that they faced similar challenges in their history, and found similar ways of meeting those challenges. It tells us that those traits they share are useful in a more universal way.

This is the case for the similarities between humans and birds. Look back at the trees of life throughout this chapter. In one sense, mammals, including humans, and birds, are pretty close. In the grand scheme of things, they are very close—both are land-dwelling vertebrates that evolved from reptiles, and much closer to each other than they are to other groups such as fish or invertebrates. Despite this, the difference between them is still vast. The last common ancestor of birds and mammals died 320 million years ago, lacked fur or feathers, had cold blood, and was essentially a reptile. More importantly, it lacked many of the traits that they now share as a result of convergent evolution.

Shared traits between two mammals, which were likely present in their common ancestor, may tell us little: it could just be a leftover that natural selection never had reason to remove. But if a human and a bird share a trait, then that similarity was not an accident. It means that those shared characteristics were successful; so successful that two different groups of animals both resorted to the same strategies. It tells us something about the 'why?' questions behind the evolution of those traits. It tells us *why* evolution proceeded the way it did, producing similar adaptations despite having different hardware with which to work.

We can then look at the similarities in the history of the different groups of animals to understand why a shared strategy was successful, and, in doing so, understand why we are the way we are. All of the similarities in this book are examples of convergent evolution, and all of them tell us as much about ourselves as they tell us about birds.

* * *

We are pretty resoundingly mammalian and have so much in common with our fellow mammals in terms of anatomy and even behaviour. What, then, are these differences that cause one to look to the birds to find similarity?

In order to see how we break the mammalian mould, we have to think about the standard mammal. Furthermore, many of the similarities between humans and birds could also be extended to other large, long-lived mammals, including livestock and the charismatic megafauna, such as elephants and tigers, that dominate our conservation efforts. It is easy to think of large, familiar animals as 'typical', but, in fact, they are all highly irregular cases, that's why they are famous. The typical mammal is much less like a cow, an elephant, or a whale, and much more like a rat.

When the asteroid that killed almost all of the dinosaurs hit, any large animals were doomed. The changes in climate caused by the impact drastically reduced the food supply, thanks to the reduced levels of heat and sunlight. Without sunlight, plants suffered, and plants are the foundation of every ecosystem: if an animal doesn't eat plants, it eats animals that eat plants. All of the energy available for life comes from the Sun, via the photosynthesis of plants (or in certain limited circumstances, in the deep ocean, from the heat of the Earth's molten interior). Once the plants were reduced, large animals could not hope to gather enough energy to go on living, and almost every single large species became extinct.

The birds, the subset of dinosaurs small enough to fly, were also small enough to survive this catastrophe, but so were many of the mammals. In the age of the dinosaurs, there was no real opportunity for large mammals to succeed. Biologists use the term 'niche' to describe where an animal fits into an ecosystem. We borrowed the word from architecture, where it refers to the arched indentations that hold statues, as you might see in an ornate church. An empty niche in a church or college wall cries out to be filled, its very form giving the obvious and distinct impression that a statue ought to be there. This makes it a good metaphor for the opportunities that exist for distinct strategies in an ecosystem. As animals evolve, the most successful ones will be those that hit on a strategy that minimizes competition and maximizes opportunity. If a region had lots of plant life and many herbivores grazing those plants, there is a big, empty opportunity for any animal that evolves the ability to eat meat. All those herbivores mean there is a lot of high-density energy wandering around and grazing, enough to keep a population of big cats or wolves well fed. Like a cathedral niche without a statue, the space is there, waiting to be filled.

If there are already lots of predators feeding on the herbivores, a new carnivore will not have the same opportunity. The addition of another carnivore might even throw the ecosystem off balance, as it could result in over-hunting of the herbivores, which would lead to a smaller breeding population of herbivores, beginning a downward spiral. There is no vacant niche.

When large dinosaurs still lived, they occupied just about all the niches available for large animals, and as a result, early mammals were small, rodent-like scavengers, mostly concerned with staying out of the way of the dinosaurs. The large mammals we see today, carnivores and herbivores alike, are the result of these early, small mammals evolving to take advantage of the niches left empty once the dinosaurs were gone and the planet had recovered from the blast.

Early mammals were mostly nocturnal. Waiting for the cover of night is a good strategy for small, vulnerable mammals trying to avoid becoming a snack for a large dinosaur.

And most species of mammals are still nocturnal. Biologists call this an evolutionary 'bottleneck'. In order to avoid the daytime predators, the mammals had to become nocturnal. This established the 'normal' state for mammals, and any modern mammals that are diurnal have had to develop the necessary adaptations for daytime living after the fall of the dinosaurs. We humans and a number of other mammal species now occupy the large, diurnal niches that the dinosaurs left empty, but there is far more space in the world's ecosystems for small scavengers than for large predators or huge herbivores. The majority of mammal species have remained small and have remained in the dark.

Being nocturnal led to other mammal norms from which we humans diverge. A nocturnal animal cannot rely on sight in the same way as a diurnal animal. Many nocturnal animals have eyes that are adapted specifically for the dark: they are larger, so that they can gather as much of the limited light as possible, and they tend not to prioritize colour vision, since colours cannot be seen well in the dark, even if you do have big eyes. Instead, nocturnal species will have receptor cells in the eyes that are specifically adapted to get as much shape information as possible, but not really able to distinguish colours. When some mammals later evolved to be diurnal, this colour-blindness was a limitation that stuck. Almost all mammals have only two types of colour-detecting cone cells in their eyes, which means that they have very limited colour vision. In the case of dogs, this means they cannot see red; their vision is more like that of a red/green colour-blind human. Only the primates are confirmed to have evolved the third colour-detecting cone cell that allows us to see a whole rainbow of colours.[16]

Nocturnal animals tend to rely very strongly on scent and hearing, instead of vision, and modern mammals still have excellent noses and ears to compensate for their colour-blindness. Again, our pet dogs are a good example. A dog may not see something very well, but he can definitely hear and smell it long before we do. The dog inherited his bad vision from his nocturnal wolf ancestor, but also retained that ancestor's good hearing and an excellent sense of smell to make up for it.

We primates are the exception in evolving a third colour detector and shifting our strategy to be mainly vision-focused. In doing so, we became much more like the birds. Since birds evolved from a ferocious and predatory subset of dinosaurs, they started off diurnal, with no need to hide in the darkness. As a result, almost all modern birds remain diurnal, and the few that are not, like many owl species, have specifically evolved towards night living (and have the large eyes to go with it). All those diurnal bird species were able to make very good use of colour vision for the entirety of their evolutionary history—and, because of this, birds have not three but four types of cone cells, and are able to see all the colours we can see, plus ultraviolet light.[17] Birds are very reliant on vision and gather most of their information about the surrounding world through this sense. Until the evolution of primates, no mammal was as dependent on vision as the birds.

Birds also have perfectly good hearing—after all, they do sing to each other, especially in the mating season—and not a bad sense of smell, but compared to mammals, they use vision for a considerable amount of their communication and understanding of the world around them. Mating dances, head bobbing, and other displays make up a large part of how birds interact. Primates switched to bird senses when they evolved their extra cone cell. Our brains allocated more neurons to processing vision—a task that requires a lot of computational power—and fewer to smell and hearing, becoming less like our fellow mammals and more like the birds. We also began to use much more visual communication. While we humans have highly developed language, based on sound and hearing, this is not very common in other primates. Chimpanzees and other apes communicate mostly through facial gestures and other movements—visual forms of communication. When humans have tried to communicate with apes, as in the case of the famous gorilla Koko, we have taught them sign language.[18] Humans, and all primates, became more like birds with regard to our sleeping patterns and our senses through convergent evolution.

This is just the first of many cases in which humans found success by being more bird-like than our fellow mammals. Once the dinosaurs had been extinct for several million years, the niches available for diurnal, larger, more active, more visual mammals had opened up. But birds were already living that way. They had a head start, found the same strategies we eventually did, and made the best of them. In some cases, doing it better than us. This is a situation that occurs over and over again, and not just between humans and birds.

There are, however, four profound aspects in which humans have parted ways with mammals and become birds without feathers. Our longevity, our breeding, our brains, and our social lives and communication are all far more like that of the birds than they are like other mammals. Each of these cases highlights a specific way that becoming human required us to stray radically from the strategies of our mammalian ancestors, and 'copy', through convergent evolution, the strategies of the birds that went before us. Each case will need a close look to see just how bird-like we have become. Convergent evolution with birds is the powerful force behind how we ended up human.

2

A LONG AND HAPPY LIFE

We humans spend an awful lot of time going on about our approaching deaths. Every form of art finds its obsession with confronting the fact that we all will die, and almost all sooner than we might rather want. Science and medicine, meanwhile, devote the lion's share of their resources to lengthening the human life and fighting the various ailments and pathogens that make our lives shorter and less pleasant. When I first started to study birds, one of the most common refrains I heard from non-scientists when I described my work was a sarcastic 'And how will that cure cancer?'. Our lives are fragile and finite, and we call them short, and moan about it.

But our lives are not short—at least, not by animal standards. The United Nations global human life expectancy in 2015 was 70.5 years. Women tend to live a bit longer than men, so theirs was 72.6 years, while the men's was 68.3. Compared to Earth, or the universe, or even the scale of human history, this can seem vanishingly short—hence all that poetry—but it is near the high end of life expectancy for vertebrates.[1]

It can be tempting to attribute our longevity not to something special about humans ourselves, but to modern medicine keeping us alive for an 'unnaturally' long time (whatever that means).

It is true that our life expectancy has risen enormously over the last century. The world average in 1900 was only about 30 years.[2] But this does not mean that the average adult's life has doubled or tripled in 100 years. Much of the change has been due to a decline in infant and childhood deaths. Throughout most of history, one of the most dangerous parts of life was the very early years of infancy and childhood, when complications at birth and diseases during vulnerable years killed a huge number of all children. When a sizeable chunk of new humans died at birth or shortly thereafter, average life expectancies were much shorter, but this was due to those very young deaths pulling the average down.[3] A human who made it through to late childhood and adolescence still had a fairly good chance of reaching a respectable age, even in ancient times. The life expectancy for an infant in ancient Rome was only about 20–30 years—but for a child who made it to 10 years of age, the expectancy rose to nearly 50 years.[4] As with many animals, the survival rate for very young babies was pretty gruesome, but those who made it through could live much longer. Modern medicine *has* dramatically increased life expectancy, but a big part of this is because it has reduced infant and childhood mortality, thanks especially to better postnatal care and vaccines, among other improvements.

Even our 50-year-old ancient Roman, dying on average 20 years younger than a healthy adult today, had an impressively long life for an animal, and especially for a mammal. We know all too well that most mammals do not last as long as we humans. This comes to us most painfully through our dogs—the animals we know best. Dogs were the very first animals we domesticated and have been bred longer than anything else to be capable of mutual understanding and communication with their human owners.[5]

Before any domestication for food or labour, we domesticated a friend, and that friend, in a cruel twist, has, at best, perhaps a quarter to a fifth of our lifespan. The record-holder for the oldest verifiable dog was an Australian cattle dog named Bluey, who died in 1939. She was 29 years old, more than double the age most dogs achieve.[6] Australian cattle dogs are a long-lived breed, but this amounts to their average lifespan being about a year longer than most other breeds—a bit over 13 years. Bluey was exceptional, and her owner, I imagine, grateful. Had he been a Roman, he might have kept his friend for his whole adult life.

Most mammals have similarly short, or even shorter, lives. A good indication of longevity for a species is the 'maximum lifespan'—the age of the oldest known member of the species at time of death. Bluey's record means that, for dogs, it would be 29. For us humans, this number can be subject to much debate, as there have been a succession of very old humans who have claimed to have attained great ages. The *Guinness Book of World Records* has always been plagued by these claims, and used to head their section on oldest humans with a warning that the subject was besmirched by 'vanity, deceit, falsehood, and deliberate fraud'.[7] Not every claim is so contentious. While there are a number of individuals with a few documents that allow them to claim ages as high as 129, the record-holder for a properly verifiable claim—supported by multiple authentic documents—is 122 years old. Jeanne Louise Calment was born in Arles, France, in 1875 and died in Arles, France, in 1997. She claimed to never have been ill, and smoked a maximum of two cigarettes a day, after meals, from age 21 to 117, when she presumably thought it was time to start taking health seriously. She enjoyed whisky in her tea. She also enjoyed attention, giving public appearances until the year of her death.[8]

Her record means that the maximum lifespan for humans is 122 years. For now.

***

The organization Human Aging Genomic Resources maintains the Animal Aging and Longevity Database (AnAge) of similar records for many species. It only includes extensively confirmed records, so a few very old cases with less than perfect data are excluded (including poor Bluey). AnAge allows us to compare the lifespans of thousands of different species by compiling the record for the oldest well-documented example of each. Given the 'vanity, deceit, falsehood, and deliberate fraud' around the record for oldest human, we can expect there to be many similarly unverifiable records for animals (especially those we like to keep as pets), so AnAge allows us to restrict our comparison only to those cases that have been rigorously verified with multiple documents. The result is a clearer picture of animal ageing and longevity.

We know intuitively that very small animals tend not to last as long as larger ones, and indeed, the three shortest-lived mammals on AnAge are three different species of shrew—all with a maximum lifespan of just barely over two years. House mice and brown rats have a maximum lifespan of about 4 years, with various gerbils, voles, rats, and hamsters all fitting in the 3–7-year range. Marsupials tend to die a bit sooner than their placental counterparts, so small wallabies have maximum lifespans around 10–12 years, in the company of mole rats, hares, and skunks. The vast majority of all mammals recorded have a maximum lifespan below 30 years, with llamas, deer, and domestic cats coming at the top end of that range. Only thirty five species of mammal exceed 50 years in the database, and almost all of these are primates or cetaceans. Horses (57 years), hippoipotamus – using the *correct* Greek plural meaning 'horses of the river' (61 years), elephants (65 years), and dugongs (73 years) are the sole exceptions. The top ten mammalian species are the only ones that make it to 74 years or above, and they are, all but one, whales.

Humans are that exception, and we are second on the list, behind only the bowhead whale. There is a confirmed bowhead who lived to the age of 211, becoming the oldest recorded mammal. Humans are next at 122, followed by the fin whale at 114 and the blue whale at 110. Nothing else has made it to 100. While the bowhead manages to nearly double our longevity, we still definitely stick out on that list of oldest mammals. Everything else at the top of the list is massive—whales, elephants, hippoipotamus: they all massively outweigh us, and none of them are particularly closely related. The next primate on the list is the gorilla at position 22, with a maximum lifespan of 60 years—less than half of ours, despite being twice as big or bigger. For mammals, size tends to roughly correlate with longevity, and we humans stick out as a major exception in this otherwise clean trend. The chimpanzee, our closest relative, is 25th on the list, lasting 59 years. They can still be a bit bigger than the average human, so we have done pretty well to double our age with respect to our relatives and to our size.

Birds are a little different. There is similarity, mostly at the low end, where a few tiny species, such as warblers and hummingbirds, have 2–5-year lifespans. The age-to-body-size correlation is less predictable in birds, so there are a few short-lived larger species, like the pied kingfisher. This African species is about 25 cm long, so not a tiny bird, but only lives about 4 years. All the same, the pied kingfisher is unusual. Most birds vastly outlive mammals of a similar size, and especially of a similar weight.

A good example of this is the mute swan, the beautiful bird common in European waterways. Mute swans are one of the heaviest flying birds, and, unsurprisingly, one of the longest-living species as well (again, the weight-to-age correlation isn't as clean for birds as for mammals, but it still works fairly well). The heaviest mute swans can get up to 15 kg, but a more common weight is roughly 10 kg. This puts it at a similar weight to the honey badger, an especially long-lived mammal for its size.[9] The honey badger has a maximum lifespan of about 30 years. A more typical 10-kg mammal is the jungle cat, with a 20-year maximum.[10] The mute swan's maximum lifespan, meanwhile, is 70 years—more than double to triple that of similarly sized mammals. One could make the argument that the comparison should be by size, rather than weight, since birds have much lighter bodies than mammals, but even then, every mammal with a lifespan over 50 years is much, much larger than a swan.

The swan is also an unusual bird—not only in the sense of how long it lives but also in terms of how *big* it is. The only birds that are any larger are flightless, and very unusual. Ratites, the group of birds that includes the ostrich and emu, are the heaviest, and are the oldest branch of birds, that diverged earliest from the other groups. The emperor penguin is also heavier than the swan, and flightless, as is the nearly flightless wild turkey—though the turkey is only barely larger. As we will see, being flightless and evolutionarily unusual changes things somewhat. So, the swan is just about the largest 'standard bird' there is.

The other long-lived birds are much smaller, comparable by weight or size with mammals such as small lemurs or marmosets that don't typically make it past 20 years. The Major Mitchell's cockatoo is the record-holding bird at 83 years—specifically, a cockatoo in the Chicago zoo named Cookie, who died in 2016. Major Mitchell's cockatoos are under half a metre long, at their biggest, with all their feathers, and weigh about half a kilogramme.[11] By comparison, the closest mammal by maximum lifespan is Baird's beaked whale, at 85 years. Baird's beaked whale is equivalent to well over twenty two Major Mitchell's cockatoos long (11.1 metres), and weighs about 22,000 cockatoos (11,000 kg).[12] Birds live much longer than mammals of a similar or much greater size.

AnAge lists twenty three bird species with maximum lifespans over 50. Most of these are similar in size to the Major Mitchell's cockatoo—species like the common raven (69 years), or the yellow-crowned parrot (56 years)—with a few heavier birds that are more like the swan—the American white pelican (54 years) and the Andean Condor (79 years). Only one bird on that list is truly heavy—the common ostrich, which barely makes it to a 50-year maximum lifespan for its 115 kg. For mammals that exceed 50 years, all but three of them are huge, weighing hundreds of kilogrammes. The smallest is the Borneo gibbon (60 years), which is a striking outlier at 8 kg, while the Baikal seal (56 years) at 60–70 kg, and humans (the obesity epidemic aside) come next.

Humans (and Borneo gibbons, for that matter) don't fit in with the mammal pattern. Judged by our weight and size, humans should be nowhere near the top of the list. Even if we went with our not-so-medically assisted lifespan of around 60, we would still be much smaller than any of the similarly aged mammals. As a bird, though, we are impressive, but fit in much better. With our 122-year maximum age, we would be the longest-lived and second-heaviest bird in the top range of the list. We are radically heavier than the very light parrots, and about one order of magnitude bigger than medium-weight swans and condors, but not by the multiple orders of magnitude by which whales and elephants outweigh us on the mammal list. Birds live a long time, and so do we.

The same is true for some very small birds. Tiny parrotlets weigh only a few grams but can live past 20—ten times the length of a similarly sized shrew. Mallard ducks can also get to around 25 years old, and the tiny storm petrel, weighing less than 30 grammes, can live beyond 30 years. This is why a pet bird can be such an unexpected commitment. A child who wants a parrot might be offered a budgie by a parent used to the 3–4-year commitment expected from a similarly sized hamster or gerbil. The budgie is not an especially long-lived bird, but it will easily triple the gerbil and live to 10 years?

AnAge is cautious and only records the most verifiable age claims, but reasonably reliable reports have macaws living over 100 years. Charlie, a female blue and yellow macaw, lives in a garden centre in south England. She allegedly hatched in 1899, and so might be over 120 years old. Her party trick is to squawk out riveting anti-Nazi slogans, leading her owner to claim she was once owned by Winston Churchill. His estate denies it, but she is old enough and it makes a better story.[13]

Our longevity, especially compared to our size, puts us squarely in the company of birds, and makes us very unusual for mammals. But why is this? Why do we live so long, even without modern medicine? And why do birds live even longer, relative to their size? The answer is complicated, but it boils down to humans and birds having a killer adaptation, something that changed the evolutionary game, and allowed us to survive and thrive longer than other animals. Something that made us each an extreme version of what it is to be an animal. The killer adaptation was different in each case, but sent us towards similar evolutionary outcomes.

* * *

When I was a child, I distinctly remember listening to some silly radio programme in my father's car—the sort of morning commuter amusement involving two guys bantering for a couple of hours. They were debating whether Batman or Superman was the superior superhero. Batman's barrister went first, giving an extended soliloquy on the nobility of Batman's approach—a physically ordinary man, without superpowers or extraterrestrial gifts, improving himself through training and inventing the gear required to become a superhero, and turning the tragedy of his parents' death into motivation to cleanse the world of evil. After a few minutes of this discourse, his rhetoric soared to a finish—Batman is a hero on his own terms, a hero who chose to be one, rather than simply accepting genetic and supernatural superiority. 'Point Batman,' said the referee. To which came the reply, 'Superman can fly.'

'Point Superman.'

Birds can fly. There is no simpler statement we can make about them (even if there are a few for which it isn't true). Overwhelmingly, birds fly, and this is a game-changing adaptation that makes their evolutionary story immediately different from that of any earthbound animal. Birds are evolutionary supermen because of it, and it shows in their longevity.

Together with their generally smaller size, flight is one of the chief reasons that birds were able to endure the mass extinction that destroyed the rest of their dinosaur brethren.[14] The usefulness of flight in such a situation is so intuitive to us that we hardly stop to think about it. Flight makes you harder to catch and kill, and so, naturally, you survive better. It is an extremely simple chain of logic from cause to consequence, and the obviousness of it means that I do not think we really consider what a wonder and a game-changer flight is, evolutionarily speaking.

For that matter, we do not think enough about how extraordinary *movement* is. As is often the case, typical human interaction with other life forms colours our understanding of what is 'normal' in the biological world. Most people's interaction with non-human life is completely dominated by plants and animals, and in particular, plants and animals that are big enough and hardy enough to live in our cities and suburbs. A rare cameo by a mushroom, mould, or yeast includes the fungi from time to time, though I suspect that for most people, the idea that the porcinis on your pizza constitute an exciting intrusion of a third kingdom of life into your day doesn't quite excite. (A gorgonzola and mushroom pizza, with its yeast dough—now that is an exciting triple threat of fungi!) Furthermore, because we are animals and acutely aware of movement and activity, the plants, numerous as they are, form a sort of background, and the 'nature' we really appreciate—go out looking for—is the animals. We like to look at an august old tree, but when walking through a mountain forest, it is the single deer glimpsed for a moment that makes the hike exciting, rather than the thousands of magnificent trees we take for granted as part of the setting.

The deer and the trees, like primates and birds, share a common ancestor, and are related forms of life, though admittedly that common ancestor is much further back than that of primates and birds. We have to go back about 1.6 billion years to find our last common ancestor with plants, and that organism was a single cell, as yet neither a plant nor an animal.[15] From that cell would eventually evolve the tree, and the deer, now so different that it is hard to think that they might be related. But they are, and their overwhelming difference from each other has a great deal to do with the deer, unlike the tree, having the gift of movement.

Movement is the peculiar habit of kingdom Animalia. Specifically, movement of the whole organism from one place to another; a whole host of plants and other organisms have evolved clever ways to move parts of the organism relative to themselves, even while the whole organism stays in one place. The most famous, the Venus flytrap, is an excellent example, with its rapid, jaw-like snapping motion seeming eerily animal as it closes around an unsuspecting fly. Substantial movement of the whole organism, though, is, for all practical purposes, unique to animals. While many microscopic organisms 'move' around—oscillating tiny hairs called *cilia* or whip-like tails called *flagella* to propel themselves—they do this moving essentially within the same place throughout their lifespan. The fastest, a bacterium called *Ovobacter propellens*, can zoom around at up to one millimetre per second, but lives and dies within the marine sediments where it is found.[16] Microbes like *Ovobacter* do move, but even in moving stay largely in the same place. Only animals have high motility—a life lived in deliberate movement, living in one place and hunting in another, or migrating with the seasons, or searching across miles for food.

If you think about each living organism as being one very long, very controlled, very complicated chemical reaction, then non-movement seems the natural course of things. A tree grows up out of soil when thousands of necessary chemical reagents combine, like a vastly more complicated version of salt crystals growing on a rock. All living things are essentially chemical reactions growing and changing over time, and when conceived this way, it makes sense that most living things do not have the high motility of animals. The chemical reaction starts, it whirs along while it has enough fuel and resources, and eventually finishes off and stops reacting—the tree dies. For a living organism to move (be it a deer or bacterium) requires a radical change to the way chemical reactions normally happen: the reaction has to become self-contained, detach itself from its environment, build systems for bringing in nutrients without fixed structures like roots, and build itself tools with which to propel itself (the deer's legs, the bacterium's flagella). The tree has it much easier—it drills its roots down into a source of nutrients (in this case, soil), and then lives in the same spot for the rest of its life. The earliest life was probably a simple outer envelope of membrane, holding chemical components inside, which would have bobbed around aimlessly in the primordial soup, with nutrients happening to crash into it from time to time. Directed, wide-ranging, purposeful movement was a big evolutionary innovation, and isn't at all the default or normal state of things.

Animals' development of purposeful, wide-ranging movement is part of what made us so different from other types of life. All that movement takes a huge amount of energy, and so animals tend to survive by eating other life—usually plants or other animals—in order to access highly concentrated stores of energy. All that movement also requires coordination, and so we had to develop brains and nervous systems to keep our bodies organized. Moving a massive, complicated chemical reaction also means you need to be able to take stores of chemicals with you and have systems for importing more nutrients in an organized fashion, and so we developed our many organs and organ systems, all hugely complex, to manage the logistics of keeping a chemical reaction going (that is, alive), even as it runs across a plain, or swims miles through open ocean.

Movement was, in its own way, a game-changing adaptation, and one that required huge shifts in order to pull it off. Those changes were worthwhile because movement gives an animal all kinds of advantages that a stationary plant does not have. Most straightforwardly, an animal can move away from a predator and towards a source of food, while a plant can do neither, relying on the luck of the draw as to whether or not enough rain will fall in the place where its seed took root, or whether it is sufficiently fast growing or well hidden enough to hold off the hungry herbivores coming to nibble it down to the ground.

So, the benefits of movement are so huge that it is worth the evolutionary investment and the effort. Why, then, is flying—just another form of movement—such a mark in the birds' favour?

A big part of it is simply that flying is itself rare. Animals emerged in the oceans, eventually developed the ability to live on land, and could only then start acquiring the adaptations to fly. Flying is, evolutionarily speaking, very difficult. An organism needs huge modifications in order to fly. You need to be very light but at the same time have enormous musculature in order to generate the necessary lift. You need the quick senses and coordination to control yourself once flying, and you need to evolve a new kind of limb—the wing—that is purpose-built for the job. All of this means that flying vertebrates (birds and bats) are rare compared with those that do not fly. There are so many variables you have to get just right to fly, and not many groups of animals can manage it.

It also gets much harder the larger an animal gets. Insect flight is not rare at all—all insects have wings, and nearly all of them fly, but their small size means that flight is an easier recipe for them than it is for birds and mammals. That balance between keeping the body light but having enough strength to provide lift gets harder and harder the heavier an animal gets, which is why flying animals top out with the 10 kg swan, while non-flying animals get up to the blue whale's 140,000 kg.

The result is that while any movement obviously makes an organism harder to catch and kill, flight puts their evasive abilities in a whole different league. An earthbound predator will find it much more difficult to catch a flying organism most of the time than a fellow earthbound one, and so, in most circumstances, will default to easier prey. Meanwhile, airborne predators are quite rare, both because flight itself is rare, and also because the size problem means that an animal large enough to eat a flying animal will have a harder time flying itself. Flying animals are rare, and big flying animals are even rarer, therefore flying keeps you far away from big, hungry animals waiting to kill you.

The advantages of flight go deeper though. It is fairly easy to see how flying can keep an individual animal alive longer. Put a mouse and a sparrow in a room with a hungry cat, and the smart money is on the sparrow lasting longer. The mouse, confined as it is to the same two dimensions as the cat, is not long for this world, while the sparrow, given an adequate supply of food, might well outlive the cat. But escaping predators is not the only reason that flight makes birds live longer. If you left the mouse and the sparrow, with enough food, in a room *without* a cat, the sparrow is still going to be very likely to outlive the mouse.

The maximum lifespans reported by AnAge are from animals that, as a rule, didn't die by being killed. Many of the longest-living representatives of different species have been zoo specimens, pets, livestock, or lived in protected reserves, and in general, they were not eaten or killed by an infectious disease. To use a phrase that biologists dislike, they died of 'old age'. Biologists dislike that phrase because it does not really mean anything. What is 'old age'? Why is 30 incredibly old for a dog and young and spry for a human? And why would being old kill an animal? You know the answer to the second question already—dying of old age is not actually dying of old age—it is dying of something going wrong. A heart valve gets worn out, or a kidney gives up, or some other organ falls apart, and the complex, interwoven set of systems that keeps you alive stops working. 'Old age' just means some part or parts of the body's internal systems have physically had enough, and come apart – failed in a way that happens not to be given a distinct disease name by our medicine. That, or the killer-by-default of so many organisms: some undiagnosed cancer, what Siddhartha Mukherjee called, in language as poetic as it is incisive, 'The Emperor of All Maladies' in his (2017) book of the same title.

The record-holding animals are the ones that made it all the way to the age at which their species inevitably falls apart. They died because something internal stopped working one day, often something we couldn't quite identify. Give a sparrow and a mouse a perfect diet and a clean, healthy living space with no predators, and they will do the same—carry on living until their organs simply give up and stop working. The difference is that the mouse is almost certainly going to keel over by 4 years old, whist the sparrow is fairly likely to make it to 10, or even 15 years or more. This difference would seem to have nothing to do with predators, as in this thought experiment we assumed there were none at all. And yet, the sparrow's flight and the mouse's predators are intimately involved in creating the sparrow's longer life.

You probably noticed a problem in the line of reasoning about 'old age'. Yes, it is true that old age itself doesn't kill anything, but instead it is the organs breaking down over time, but that still doesn't answer the basic question: why do organs break down when they do, and why do a human's organs last longer than a dog's, and a sparrow's longer than a mouse? And why do organs break down at all, anyway? If the human body can repair skin and keep it springy when you are 25, why does it go wrinkly and repairs slow down when you are 85?

Animal bodies have amazing abilities to repair themselves. Our bones heal, our skin replenishes itself, bruises clear, and some organs, like the liver, can even re-grow after large portions have been destroyed or removed. Our repair processes are always working to keep our organs and systems refreshed and undamaged—so much so that for humans, almost all of our cells are replaced, bit by bit, about every 10 to 15 years. (There are exceptions: most of our neurons, some cells in our eyes, and our tooth enamel are with us for life.) With all of these repair systems in place, and most of our cells being regularly replaced, it should seem a bit odd that our organs do eventually break down. We have all the tools to repair ourselves, so in theory, one might think, we could keep repairing and replacing things forever. Unfortunately, though, that doesn't tend to work, for two different but related reasons.

The first is pretty simple to understand. All living organisms have a very important job: maintaining *homeostasis*. Homeostasis comes from the Greek for 'staying the same', and maintaining homeostasis is one of the most important parts of staying alive. When biologists use this term, they tend to use it to refer to all of the different factors that have to be controlled to keep a living organism healthy and thriving: the pH of tissues, the balance of chemicals, the function of organs, and countless other factors have to be kept balanced and functioning. This all takes a lot of energy, and the body consumes energy for its various systems that do things like correcting pH levels or staying warm. Usually, when things go wrong, the body can spend a little bit of energy to put them right. As we grow older, and more things go wrong, a particular combination of problems can result in a situation that the body cannot fix. A very simple example of this is healing wounds. Ordinarily, if you get a cut, the body consumes some resources to repair the wound and all is well. If you have thinned blood or a reduction in the number of *platelets*, small cells that help the healing process, then the combination of that problem with the problem of your new cut makes it much harder to repair the wound. As we age, and problems hit at the same time and in new combinations, eventually, the complicated web of systems holding us in homeostasis fails. Homeostasis ends, or in other words, we die.

The problems come faster and in more combinations as we age because of the second reason we can't keep repairing ourselves forever. In order to create new cells in the body, our cells have to reproduce, dividing into new cells. Each time they do this, the DNA in their nucleus, which controls the cell, and contains the instructions for all of its behaviours, gets a little bit shorter. The reason for this is a bit involved, but put simply, DNA encodes genetic information as a series of nitrogenous bases—molecules of four different kinds (abbreviated A, C, G, and T) that link together to form DNA strands. When the DNA is replicated, the enzyme that copies DNA (DNA polymerase), together with other enzymes that form the replication machinery, needs to grab on to the existing strand of DNA in order to make the new copy. The enzymes then run along the DNA like a scanner, reading the DNA bases as it goes and copying them onto the new strand. Unfortunately, the enzymes are not able to copy the section they first grab on to. This section is only a few bases long, and usually at one end of the DNA strand, but each time the DNA is copied, the strand loses that little bit of DNA at one end.

We have evolved a solution to this problem, called *telomeres*. Telomeres form a 'cap' on the DNA, several hundred bases long, that doesn't code for any cellular behaviour—it is just extra DNA that is there to absorb the 'loss' each time the DNA is copied. As a result, instead of losing sections of our important genes, we just lose a bit of telomere each time. Eventually, though, after enough cellular divisions, the telomeres get completely used up, and the important part of the DNA starts to get shorter with each division.[17] With its genes damaged, a cell becomes less effective at its job. Over time, as more and more of the body's cells reach the point where their telomeres have been worn away and start to malfunction, an individual develops increasing problems that undermine his or her homeostasis.

So, a combination of factors, including how quickly cells divide, how long the telomeres are, and how much wear and tear an animal takes, will ultimately determine how long it can live before things break down. Why, though, do these factors differ so much from one species to another? Wouldn't it make sense for every species to have a slow cell division cycle, with long telomeres, and excellent repair systems, so that they could stay alive as long as possible? Why does a sparrow manage to maintain its homeostasis longer than the mouse?

Think back to the very origin of life, and to the earliest organisms. These were single cells, or even not-quite-cells, replicating in the primordial soup. They were not very complicated compared to modern life, and, for the most part, they did not live very long. It is also hard to define how long the 'life' of an asexual single-celled organism is. After all, when it reproduces, it splits into two daughter cells, and the original cell is gone. Reproduction is a kind of death, and by definition, you can only do it once. There is a philosophical question that goes deeper here, and actually touches the core of evolutionary biology itself, namely, if you split in two to create offspring, are they really a separate organism from yourself, and do you ever really die? But leave that aside. For our purposes, the important thing to note here is that the default starting state for life on Earth is a short lifespan, with only one cycle of reproduction.

That is the starting point for all the organisms that come later. Evolution is not a force with foresight; it cannot predict what tools and abilities might be nice for an organism to have. Natural selection just allows the most successful organisms to survive, and lets the others die. What this means is that if the starting point is a short lifespan with only one chance to reproduce, the forces of natural selection and evolution will, most of the time, stick with that model, and produce organisms that are better and better at making the most of that short lifespan to produce as many offspring as possible from that one reproduction.

In order for natural selection to result in a fundamental change to that model, one of two things needs to happen. On the one hand, if the existing model is doing a really bad job, then either a new model is hit upon, or the species will die off. Alternatively, even if things are going well, if a new strategy arises by chance, through random mutation, that is radically better than any other, it can become so successful that it makes the old method seem a failure. All of these changes have to happen by chance, at first. Evolution cannot pre-plan a change. But it can, and does, quickly recognize a new strategy that is much better than an old one.

Living longer and reproducing more times are now a very common strategy among a variety of different types of life. Many plants are perennials, and can live far longer than we humans, and reproduce every year. Most animals reproduce many times in a life that spans several years to decades. Obviously, this change in strategy has its advantages. In fact, because it requires several big evolutionary changes from living a short life with one big reproductive event, we know it has to have some advantages, or it never would have arisen in so many species.

It is not a perfect victory for long life, though. We still have lots of annual plants, like tomatoes, that grow from a seed to full size, fruit, produce the next generation of seeds, and die, all in one year. And some animals, like octopodes or salmon, stick with a single reproduction in their entire lives. They hatch, grow to adult size, mate, invest all their energy into that one mating event, and die shortly after reproducing. This works for the salmon and the octopodes because they are making a wager. In both cases, the animal invests every last shred of energy it has into producing the biggest, healthiest batch of eggs it can. A common octopus can lay thousands of eggs in her single clutch, and spends months without eating, guarding and aerating her eggs to make sure they are successful[18] (Figure 10). She is betting that by giving everything that she has—including her life—for the success of her eggs, she will produce more successful adult offspring than if she were to invest less into several small broods over many years. Salmon, which have to invest vast stores of energy to swim upriver to their freshwater spawning sites to breed, are making a similar gamble—they

Fig. 10 Octopodes reproduce a single time, at the end of their life. The mother octopus forgoes all food to meticulously protect and oxygenate her eggs as she wastes away, dying shortly after they hatch. These r-selected organisms literally put 'all their eggs in one basket'.

are literally putting all of their eggs into one 'basket'—one shot at producing as many offspring as possible, consuming everything that they have in the process. Gardeners like tomatoes for the same reason: a healthy tomato plant produces huge numbers of fruit for its size, using all of its energy stores for its one shot at reproduction.

This strategy is what biologists call *r selection*.[19] Here, 'r' is the term used in population dynamics for the growth rate, and r-selected species have fast growth rates. They grow from birth to sexual maturity quickly, and they have large broods of offspring to which they do not give very much individual care (after they hatch or are born). Salmon and octopodes (and tomatoes) are examples of organisms that are r selected in the extreme—one great big breeding event per lifetime, lots and lots of offspring that are basically on their own once they hatch. In an r selected organism, lifespans are short, because living any longer doesn't come with any advantages. If you are a salmon with a gene that allows you to live an extra year after your spawning, you have no advantage as far as natural selection is concerned over the salmon who dies right away—you reproduce once, they reproduce once, and your extra year of retirement does nothing at all to help you produce more successful, adult offspring. You are probably so exhausted from the spawning migration anyway that even if you could breed again, you'd be eaten by a bear before you got the chance. The gene pool of salmon will not be taken over by your gene for long life since it confers no reproductive advantage at all. Similarly, there is no evolutionary pressure for tomatoes or octopodes to live any longer. A mutation that got rid of an aspect of their quick degeneration would not actually improve their reproductive success, so there is no evolutionary pressure to push their lifespans longer.

The opposite of r selection is *K selection*. 'K' is the term for 'carrying capacity', or the number of organisms an environment can successfully support—an important value for a species that lives a long time. K selected organisms live long lives and reproduce many times, with smaller broods, and in many, but not all cases, spend a lot of effort and time looking after their juvenile young, sometimes for years. Humans, other primates and large mammals, and, of course, birds, are all strongly K selected, and live a long time, with multiple matings. A K selected organism has a very good reason to live longer—the longer you live, the more times you can reproduce, and the more successful you can be in the competition to have a larger number of healthy offspring. If an average duck lives 20 years and mates 12 times, then a duck with a mutation that allows it to live 25 years and mate 15 times will, on average, have perhaps 20 per cent more offspring. And so, over time, more and more ducks will have that gene, and lifespan will get longer. This process becomes a virtuous cycle, and lives get longer and longer, as long as the payoff is there. Similarly, unlike the r selected organisms, if a K selected organism has a mutation that causes it to lose some element of ageing, it can do very well indeed, and that mutation will probably spread, as it means the organism can keep having more broods for longer.

K selected organisms are not immortal, though, so what prevents the virtuous cycle from continuing forever, producing longer and longer lives every generation for K selected animals? There are a number of elements, but all of them come down to trade-offs, and whether living another year is 'worth it' for natural selection. For example, take any mammal, a dog, for example. Dogs can reproduce for most of their adult lives, so having a longer adult life means more puppies. But only up to a point! As any mothers reading will know, pregnancy and birth take a serious toll on a body. It has risks associated, and death by childbirth used to be among the most common sources of female mortality. Dogs have childbirth quite a bit easier than humans, but it is still an energetically intense thing to do and causes more bodily damage, the more litters a dog has. There will be a limit to how many successful pregnancies a dog can have before the physical damage is such that the next litter is too small, or too unhealthy, or too likely to kill the mother. At that point, living another year doesn't matter any more, because it will not likely result in any more puppies. The virtuous cycle stops.

These kinds of intrinsic reasons are important, but the big one is predation. And here is the great double virtue of longevity: the harder it is for predators to kill you, the longer your natural lifespan. Remember, this is not just because you outrun the predator—an animal that is hard for a predator to kill will live longer than one that is easily killed, even if both of them are kept safely in a predator-free enclosure. This is the same trade-off as the pregnant dog earlier. A mouse is very easy for a predator to catch. As a result, in any given year, it is likely that the mouse will be caught, killed, and thus no longer able to breed. Therefore, the mouse needs to be r selected—it has to make the most of what little time it has and produce a large number of offspring quickly. Evolving to live longer than a few years is not worth it because, by the time the mouse is a few years old, its longer lifespan won't matter—the cat's already eaten it.

And so, flight is the magic bullet. Flight, with its unparalleled ability to keep an animal out of harm's way, is the ultimate K selector. A bird is unlikely to be eaten in any given year, because it is able to fly away. This means that living until next year is worth it—in all likelihood, the bird will still be alive, and producing more eggs and offspring. And the next year, and the year after that. Flight is the secret ingredient to birds' long lives because it makes every extra year worth it. And flight is not just for the birds—for what it's worth, bats, the *other* birds without feathers, have lifespans wildly out of proportion for mammals, but perfectly ordinary for birds. Flight is an extreme way to move and evade predators, and so leads to an extreme version of K selection.

Parrots, like the record-breaking Major Mitchell's cockatoo and macaw, can produce healthy clutches of eggs for decade after decade after decade—flying away and outsmarting the animals that might kill a parrot-sized possum stuck down below. And with all that time, they don't have to mate every year to make the most of their time. They can rest and recover from egg laying, nest sitting, and chick feeding. They have time, and so they get more time, because their time is worth it. So worth it, in fact, that they can outlive us.

Birds, like humans, break the pattern that links size with longevity. All of us vastly outlive our size and weight by comparison to other animals. For the birds, flight is the reason—the fundamental and defining feature that extends their lives. It is also the reason that for birds, the very largest species are not, in fact, the longest lived. Ostriches cannot fly, and so do not get that extra boost of longevity and are beaten by the flying swans and cockatoos.

Obviously, humans do not fly. If flight is the birds' secret weapon for longevity, what is ours?

Well, we humans do fly. Or at least we do now. It took us many tens of thousands of years to pull it off, but with a variety of assistive technologies, we fly, and we fly faster than any other animal on Earth. Flight is not our secret weapon, just a consequence of it.

Brains. Our brains are our flight. Humans found, in big, powerful, complex brains, a super-specialty that makes us wildly successful and very hard to kill, just as birds did with flight. What our big brains share with flight is that they empower us to be flexible. A bird can escape a sticky situation by running away in a direction few other vertebrates can: up. And humans get out of our own sticky situations with wily intelligence—or keep ourselves out of such quandaries in the first place. Our intelligence gave us tools and rules, and ultimately zeppelins and jetliners. It is an extreme animal trait in the same way that flying is.

Humans and birds share our happily long lifespans thanks to our defining, 'secret weapon' survival adaptations of brains and flight, respectively. These shot us out ahead of other animals in success and longevity. And by a brilliant symmetry, as our big brains led us ultimately to achieve birds' flight, their flight led them ultimately to share our big brains.

3

BIRD BRAINS

Bird brains. My least favourite cliché. A slur among slurs, and perhaps the most unfair preconceived notion the average person has about the animal kingdom. The English language is stacked with idioms of birds, birdly turns of phrase, from 'lovely weather for ducks', to 'don't count your chickens', to 'a bird in the hand is worth two in the bush'. In any case, these innocent, colourful phrases do not typically mean much, correctly or otherwise, for our understanding of actual animals, except for 'bird brains'. Of all the things one could say about birds, of all the comparisons one could make to stupidity, this is perhaps the most unjustified common expression: the idea that birds' brains are unimpressive or stupid.

As you will no doubt have imagined from the above, birds have rather impressive brains. Birds, as a group, are very intelligent as animals go, and some birds really do hold their own among the cleverest animals on Earth.[1] The slur seems instead to be predicated on the incorrect assumption that brain size is what determines brain ability. Looking at a chicken or a wren, the average observer sees what, compared to us as self-important humans, is a small head that can only contain a small brain. Indeed, birds do have quite small brains, which makes it all the more impressive that these small packages contain some of the most powerful animal minds. To make up for their smaller size, birds' brains are efficient, cramming into very small spaces computing power more in line with much larger brain volumes. The largest bird, the ostrich, has a brain about the size of a large plum, and is very far from being the smartest bird, with parrots and crows having smaller brains still.[2] This spatial efficiency of birds' brains is only the beginning of their bragging rights, and birds themselves are one of the stronger arguments against using straight brain size to measure intelligence.

Still, discussions of animal brains often begin with size, and though birds throw off the general trend, just as they do for longevity, there is still real value in investigating what brain sizes look like across species.[3] So, what are the largest animal brains? The undisputed champion in this area is the sperm whale, with the largest brain on Earth, at about 8 kilogrammes,[4] but ranging higher.[5] The orca's brain gets pretty close to that as well. Unsurprisingly, a variety of other whales and dolphins also have very large brains, in keeping with their overall large size. Dolphins' brains are generally around the 3 kilogrammes mark,[6] and the biggest brains on land are those of elephants, weighing in at roughly 5 kilogrammes. Humans' brains, which come in the vicinity of 1.3 kilogrammes,[7] seem puny by comparison, but as with our longevity, our body size makes us the odd one out. True, there are bigger brains than ours out there, but only in animals that vastly outweigh us, often by orders of magnitude. That 8 kilogrammes sperm whale brain, about five times heavier than ours, sits in an animal that can weigh 57,000 kilogrammes, several hundred times our size.[8] Looking at animals closer to our stature, we break the pattern for brain size rather substantially. Our brains are about three times the size of those of our chimpanzee cousins,[9] despite roughly similar body sizes; and even that is an unfair comparison, because all primates have rather large brains for their size.[10] Those of other mammals in our weight class, such as deer[11] or seals, are smaller still. No prizes for guessing that the smallest vertebrate brains come in tiny animals, like shrews and mice among mammals,[12] and even smaller are those of tiny reptiles and fish.

Total brain size is not wholly useless as an indicator of intelligence (whales, elephants, and humans *are* very intelligent animals), but it is clearly not the whole story. We are, definitely, smarter than sperm whales and dolphins, despite their vastly bigger brains. This might be fine if humans were the only exception to the trend, one animal with a highly efficient brain for whom size is not the main indicator of intelligence, but this is not the case. Everywhere we look, we find relatively small brains doing impressive work, often better than larger ones. Tamarins, a type of diminutive primate, come with a primate brain and all of its problem-solving, pattern-recognizing, social relationship-managing skills. This brain is probably (this sort of comparison is always fraught, hence my weasel words) more capable than the physically larger brain of, say, a deer; or for that matter, the similarly sized brain of a weasel. Besides, we have already noted that our intelligent birds have very small brains. How, then, can we make meaningful judgements about brains across wildly diverse animals?

My comparison of the body size of whales and humans just now should have given a clue as to a better way. The brain-to-body mass ratio, that is, the size of the brain relative to the size of its body, can offer a slightly more nuanced view. The argument here goes that it is not the absolute size of the brain that counts, but how much of your total body weight is brain, or, in a sense, how much of your mass you have invested in brain power, relative to other needs, like musculature and digestive organs.[13]

Looking at brain-to-body ratio, we start to see some patterns that make sense. An elephant, the brain size champion on land, has a ratio of about 1:560, that is, a body weight 560 times heavier than its brain alone. A horse, with a smaller body and smaller brain, comes in at a slightly smaller 1:600, a bit less brain per body, in line with our perception that elephants are probably a bit smarter than horses. Humans, meanwhile, have a much larger ratio of 1:40: a much bigger brain relative to our body, and a radically different number when compared to the elephant, in keeping with our massive self-regard for our own intelligence. Well done, us. So far, this metric seems to be working. It is also a very good metric for the thesis of this book. Many small birds have ratios up near 1:10 or 1:12, higher even than we humans. So, job done, then, right? Birds and humans, among all their other similarities, have very high brain-to-body ratios, and birds even beat humans, to top things off! Well, no. True, most birds have a very large brain-to-body ratio, much more similar to that of humans than we are to other mammals. But that alone should set off some alarm bells—birds are very smart, but none of them are on a level with humans. Looking a bit closer, we see that this metric also starts to throw up some weird comparisons. Though horses and elephants seemed about right compared to each other, domestic dogs and cats come in much higher, around 1:125 and 1:100, respectively. It is always hard comparing intelligence across big mammals, but domestic dogs' 1:125 seems wildly out of line with the elephants' 1:560. Moreover, lions are all the way down with elephants at 1:550, when really their behaviour and intelligence should be looking most similar to other cats. And mice, for goodness sake, are exactly tied with humans at 1:40![14]

Obviously, there is more to this ratio than intelligence, and you have probably already seen it. The bigger an animal is, the lower, generally, its brain-to-body ratio. So, the very comparable cat and lion are 1:100 and 1:550 respectively, but the lion's whole body is vastly bigger. This makes sense—cats and lions have reasonably similar behaviours, and need reasonably similar brainpower to control those behaviours. The larger size of the lion's muscles, organs, and overall body doesn't make controlling them more complicated, so its brain doesn't need to scale up exactly with its body compared to the cat. Conversely, in small animals, even if behaviours are rather simple, brains can only be miniaturized so far. So, a mouse's brain has to be small to fit into its small body, but ends up with a much larger ratio simply because the brain couldn't be shrunk much further. This can be seen very obviously by looking at some species of ants, with absolutely tiny brains, but bodies so small that the brain makes up 1/7th of their total weight.[15] In the case of birds, flight, as always, comes into the equation as well. Birds are, overall, very light for their size, thanks to hollow bones and thin limbs that make them light enough to fly. Their lightweight bodies make any brain appear heavier by comparison. The only really meaningful comparisons of brain-to-body mass ratio happen between similar animals of roughly similar mass and stature.[16] The comparison of the elephant and the hippopotamus is revealing: compared to the elephant's ratio of 1:560, the hippopotamus comes in at 1:2800. The elephant is smarter, and, for once, I am pretty confident of that.

As an aside, the animal with the smallest brain-to-body mass ratio is called the bony-eared assfish.[17] In case you ever thought nature dealt *you* a crummy hand.

* * *

Brain size doesn't work. The brain-to-body mass ratio doesn't work. Does anything? Not really. The truth is, with the incredible diversity of body shape, size, environmental needs, and other characteristics that affect animals, a clean, reliable physical measure for intelligence does not seem to be forthcoming. One factor involved is that not all brains distribute their weight in the same ways. Take dolphins, for example. Next to humans, they are perhaps the most worrying comparison when it comes to our own egos. Dolphins aren't all that much bigger than we are, but they have much bigger brains. Mammal brains, though, are made up of a number of different types of cells. Neurons, the typical 'brain cells' most people think of, are the information-carrying ones that do the work of computing, reacting, and thinking. Other cells, called glia, play a supporting role, providing structure, carrying out repairs, and myelinating the axons of the neurons. Dolphin brains contain a high proportion of glia, about twice the number, relative to neurons, as humans do, adding mass and boosting brain size, but seemingly not increasing computing power or intelligence.[18,19]

Birds also show a difference in their brain structure. Their neurons are smaller than those of mammals and packed in more densely. This is at the heart of the increased efficiency of their brains, and the reason why their very small brains buck the trend and can compete with the larger brains of mammals and even primates. The result is that, compared with mammals, very small birds, with similarly small brains, will have many more neurons than similarly sized mammals. Small songbirds can weigh in at barely a tenth the weight of a common mouse, but sport more than double the number of neurons. Meanwhile, some of the heaviest bird brains, which are found in the macaws, the big, colourful South American parrots, weigh in at perhaps 20–25 grammes.[20] This is a bit bigger than the brain of a common European rabbit. Yet while the rabbit has about half a billion neurons[21] in its whole body, the macaw can have over three billion in its brain alone,[22] a number more in line with much larger giraffes and baboons.[23] There are really only four types of animal that get into the billions of neurons: whales, large mammals such as elephants and seals, primates, and brainy birds (parrots and crows). Once again, as with longevity, we see diminutive primates and relatively tiny birds soaring with the largest animals on the planet when it comes to neuron numbers.

Humans have 13–16 billion neurons in the average brain—a few multiples of a macaw but in a similar neighbourhood. If we packed our neurons as densely as the macaw does, though, we could have as many as 160 billion. This seems like a great evolutionary strategy for the macaw (and all the other birds), so why do humans and other mammals not pack in as many neurons as the birds?

By now, you can probably say it with me: flight.

Shrinking neurons is evolutionarily hard. On the surface, it is hard the way all structural changes are evolutionarily hard. If a biological solution is working, changing it risks upsetting the apple cart and causing more damage than benefit. Most biological systems are very finely tuned, so the majority of changes are damaging rather than helpful. In order for a species to pull off a structural change, there needs to be a strong evolutionary reason to do it. Otherwise, the status quo is safer. Mammalian brains work—they work very well, in fact, and modifying them with a big, global change, such as shrinking every neuron, is a very risky move. Shrinking neurons is even harder than usual though, because, in order to pull it off, you have to shrink the species' whole genome. The genome, remember, is all the genetic information in a given cell. Every cell in the body (minus a few exceptions like blood cells) has a full copy of the whole genome. DNA is an astonishingly compact way of storing information, but the genome is an enormous amount of information, and the chromosomes of DNA in each cell do take up physical space. You can only shrink a cell so far before you reach a minimum limit to how much room you need in each cell to hold the chromosomes and organelles that keep the cell alive. Birds have shrunk their neurons so small that they have had also to shrink their whole genome to pull it off.[24]

Think about that! These are the instructions that themselves determine how the body assembles itself—including information like how big to make the neurons! The pressure to become smaller and lighter was so strong in birds that they had to literally optimize the file size of their genome just to fit it into each cell.

Flight is worth it. Just like the hollow, fragile bones that make birds physically less robust but keep them light enough to fly, the shrunken neurons and the smaller genome that allows for them are in service of keeping birds airborne, but that is an enormous evolutionary prize. Mammals, which, other than bats, do not have any hope of reaping the benefits of flight, simply do not have the huge evolutionary pressure required to make all of that change and all of that genome editing worth it. Bats, meanwhile, also have a smaller genome, for the same reason.[25] No surprises there.

Of course, having smaller neurons and a lighter brain and head had to come before flight, since it is part of what makes flight possible. Indeed, evidence suggests that the groups of dinosaurs that would over time become modern birds were shrinking their genome and cell sizes for millions of years before they achieved lift-off. Birds' flight required densely packed brains, but it didn't make those early flying dinosaurs radically more intelligent than their earthbound counterparts. They would have had similar numbers of neurons packed into brains that happened to be smaller, allowing flight to evolve. It was only later that the small, densely packed neurons were at the ready to make highly intelligent birds possible, and keep them flying.[26]

* * *

By this point, some of my colleagues in the animal behaviour world will be mad at me. I have been going on for pages about intelligence, one of the most controversial topics, and even words, in biology, and I haven't yet defined what I mean by that word. *Mea culpa.*

Part of the reason is that I don't want to get bogged down in the professional squabbles about which kinds of behaviour and abilities are rightly called 'intelligent'. We all have some intuitive sense of what we mean when we use the word, and I think that intuitive sense is really and truly meaningful, if hard to define. Another part of the reason is, as you might have guessed from my noncommittal comparisons of animal intelligence, it can be quite hard to pin down quantitative comparisons between animals when it comes to intelligence, and it is often safer to talk only in generalities. Then, even if you are not right, at least you are not wrong.

All that said, my angry colleagues are right. We need to define intelligence in order to talk about it productively, not least because, intuitive sense aside, quite a lot of people get quite a lot wrong about what animal intelligence is, and how we can identify and talk about it.

* * *

In most years, I teach a class on animal intelligence. It is one of the most enjoyable lessons I give to Biology undergraduates, both because it is the closest topic to my own laboratory research that I actually teach, and because it produces such interesting discussion among the students. But while the overall conversation is always different—often with the year's cohort introducing me to a new study or recent development in the field of which I am not yet aware—there is one conversation that happens every year like clockwork. Before the students and I dig into the real meat of the conversation, I like to make sure we are talking about the same things, so I always ask them, 'What is intelligence?'. Without fail, someone will chime in with an attractive definition:

'I think intelligence is, sort of, doing the right thing, or right behaviour, at the right time and in the right way . . .' one will offer.

Not bad, and certainly intuitive, but not right. So, I respond:

'So, if a flowering plant opens its flowers during the right time of day for the insects that pollinate it, does that make it intelligent?'

'Yes—I think all organisms are intelligent in their own way. That plant is not intelligent in the same way a human is, but is as intelligent in its own way if it can successfully do that.'

The definition is not always the same, nor my example, but that conclusion, of the universality of intelligence in organisms whose behaviour 'works' as intended, is always—without fail—the first-offered theory of intelligence. I understand why this is the case. School children are rightly taught to wonder at the beauty and cleverness of nature. A good biology teacher should remark on the grand 'fittedness' of flowers opening at just the right time for best pollination. This then gets alloyed to another common schoolyard affirmation: 'everyone is brilliant (special, worthy, what have you) in their own special way' (or words to that effect) *et voilà*, my green young biology undergraduates produce 'every species is intelligent in its own special way'.

I cannot blame them—defining intelligence is not an easy task. In a casual sense, there isn't anything wrong with their attempt. This is because I suspect 'intelligence' is standing in for 'well-adaptedness'. And then, there is some truth to it. Any species continuing to survive has those tools necessary to do so. Having such tools is useful for an organism, and usually does result in most organisms doing 'the right thing at the right time', but in biology, we call those tools—claws, mothering behaviour, flower timings, migration paths—adaptations, not intelligence. And while at the dinner table, many might reasonably call them 'clever', and almost as reasonably call them 'intelligent', it does not take any intelligence (in a biological sense) to be well-adapted.

This may seem a bit of a quibble—even a pointless exercise in language pedantry, but the casual confusion of these two is, to me, regrettable. Importantly, if all we mean by intelligence is well-adaptedness, then we ignore the interaction between the two: we fail to see intelligence, properly defined, as one of myriad adaptations that an organism may manifest, but also as an adaptation that redefines how an organism can be successful. Intelligence is a more powerful tool of survival and more dramatic force for change than any other characteristic of life. Intelligence enables change—rapid change. It is about invention and innovation, analysis, and modification.

However we decide to define intelligence—precisely in the lab or intuitively in the pub—it is an impressive characteristic. When thinking about animal intelligence, most people's minds likely jump to chimpanzees or parrots, and we are comfortably reassured by our own position as humans. In some historical contexts, we were thought of as the *only* intelligent species—with those powers of innovation and analysis being the defining features of humanity: John Locke contended, of our fellow animals: 'brutes abstract not'.[27] More often, we give some credit to our fellow vertebrates and call ourselves the *most* intelligent species. Either way, we recognize in ourselves a difference in kind or in degree of intelligence.

Ultimately, building a definition for intelligence is not all that hard. What makes intelligence different from the well-adaptedness my students first offer is the inventiveness of it. It is easy for a well-adapted organism, like a flower, to do something useful, like bloom, if that process is built into its DNA, and impossible for the organism to stop, wilfully start, or modify. Many useful things that organisms do are like water freezing into ice: one chemical or physical process leads to another, and the system uses this to produce some output, every time.

A good example of this is 'bud break' in grape vines.[28] In the winter, grape vines go dormant and brown, before beginning to grow again from dormant buds in the spring. This happens just on time, because, as the air around the vine warms, the pressure exerted on the buds by water in the vine changes, causing them to swell and then 'break', beginning the new year's growth cycle. It is a system that works well, and is undoubtedly well adapted. But it is not intelligent, it is automatic.

Intelligence is the ability to invent a new solution to a problem, a solution that does not rely on the built-in solutions that evolution has provided. It is having the capacity to create a behaviour that is not instinctual but novel. It is the ability to innovate and to analyse, and do so whenever the on-board biological pre-programming confronts a situation for which there is no automatic response. It is, fundamentally, the ability to change strategy within a lifetime.

* * *

Intelligence is an incredible adaptation. It is humanity's silver bullet, our game-changing adaptation that led us to be so different from other mammals. It is a critical part of why we are so successful and live so long.

Imagine an animal with no intelligence at all, just well-adapted innate behaviours. It is an automaton. It can create no new or different behaviours, but is perfectly evolved to live successfully in its environment. Like a computer, it has a complex suite of programs and responses to a very large number of inputs, so that it can respond perfectly to those inputs, but not to any inputs that are not pre-programmed. As long as nothing in its environment changes, the automaton will do well—its behaviours, perfectly tuned by aeons of evolution, will respond correctly to the large but limited set of circumstances with which it is prepared to deal. If something changes, however, it is in trouble. With no in-built response, the only way for the animal to modify its behaviour is through evolution—which is death. Literally death—evolution happens when some animals with a good set of adaptations for their environment survive, and those without do not. The species of automata can improve their response to a new circumstance, but only by countless numbers of them dying, having failed to reproduce, until a few lucky ones have mutations that change their normal behaviour in a positive way. That can enable the species to survive but does nothing for the individuals that die in the process.

You might think there is no such animal, but some are not very far off—particularly in the oceans. Sponges and corals have more or less this problem. They are animals, but they lack a brain and have very limited sets of precise and essentially automatic responses to different threats and circumstances.[29] A coral polyp under physical threat can retract into its skeleton or sting. It can eat by grabbing particulate food and pushing it to its mouth.[30,31] It can expel its photosynthetic microorganisms if the temperature gets too warm, and some species can bail out of their skeletons if under extreme chemical threat, though this is usually a death sentence. And that's about it. Whole reefs can bleach when a sudden temperature or chemical change occurs. The corals cannot change their behaviour to try a new strategy—they can only live or die with the strategies they have. (And except for the very risky skeleton bail-out strategy, they also can't move, which makes things even harder!)

Most animals, though, have a brain, even a small one, and have some degree of flexibility. But it varies widely. Small insects, other invertebrates like worms, and even many simple fish do not have very much brain power with which to analyse a situation and invent new behaviour. They have more options than the corals, mostly creative directions to try a retreat, but not much. The range of animal intelligence is huge, and we humans are really and truly special in this regard.

We are the smartest animals on the planet. That is perhaps the intelligence comparison I am most confident in, and most comfortable putting down on paper. But it is also pretty obvious. For an ape that evolved for tribal living, outdoors, in warm African valleys, not a million years later, to also be occasionally living in metal canisters orbiting the planet is a level of inventive flexibility that has no equal on this planet by a long shot.

This is how intelligence became our flight. It put us on an entirely different metaphorical 'plane' from other animals, much as birds' flight put them on a different physical plane. We could build weapons, shields, clothing, and shelters that made us hardier still and harder to kill, until ultimately there came to be very little on Earth that could regularly cut our lives short. Our lives grew longer, like those of birds, and no matter what a long life threw at us, no matter how much circumstances changed, at least some of us could adapt and overcome.

Long-living animals need to be smart, precisely because they are not disposable. Think back to r-selected species from Chapter 2. If you are a short-lived, quickly reproducing mouse, your life is more or less disposable—at least, evolutionarily speaking. You reach maturity quickly and breed fast, creating generations in quick succession, and ensuring that, by barely a couple years of age, you might already have dozens of grandchildren, all keeping some of your genetic material alive in the world. If you make a mistake, or don't have the flexibility to dodge a fatal encounter (like a cat), then, while it is the end for you, your genes are not much worse off—you have already succeeded by putting your energy and ability into creating dozens of offspring. A very K-selected elephant, by contrast, needs to be clever. The elephant will take decades in order to produce a few offspring. Decades is the sort of timescale on which everything can change. Droughts, fires, human activity, storms, and many other changes and catastrophes can all make the same range of savannah vary considerably over the course of many years. Sometimes a boom of predators, sometimes a famine during drought. The world constantly changing. An elephant has to be smart enough to stay alive through all of it just to get to its first breeding.

Long-living birds have the same pressure. A crow that can live for 50 years takes a while to reach sexual maturity, and once it does, it has decades of productive life ahead in which to produce offspring. Flight helped make the crow long-lived, but now it needs to be flexible and creative to stay successful—and stay alive—through as much of its potential lifespan as possible. We came at it from different angles, but humans and birds have both ended up in a virtuous cycle: we both live a long time, and so need to be intelligent; we are both intelligent, and this helps us live longer. The two traits act on each other, and natural selection pushes both traits further with the help of the other. Once caught in this virtuous cycle, both birds and humans get pushed towards the same extreme, becoming more like each other and less like their near relatives, even though they started so far apart on the evolutionary tree.

Longevity and intelligence interact in another way as well, another feature that humans share with many birds. Although brain size is not a perfect guide to intelligence, it is fairly clear that, overall, a bigger brain compared to an otherwise very similar animal will usually mean a more intelligent beast. Growing bigger brains takes time and energy, and this is particularly evident in the early life development of intelligent animals. Humans are, again, the most extreme example in this regard. Our babies are effectively born prematurely compared to every other mammal. Human babies cannot do anything for themselves, even compared to our closest relatives. Infant chimpanzees can cling to their mothers from almost the moment of birth, holding on to her hair so she can go about her business with both hands.[32,33] More distant mammals have it even easier, with many large and reasonably intelligent animals like horses or deer being able to walk or stand within minutes to days of birth, compared to humans' months or years to acquire any useful behaviours.

This is because humans are born with an underdeveloped brain. Our labour and birthing are, without comparison, the most dangerous and difficult among mammals (and really all vertebrates). The duration, pain, and risk associated with childbirth are something no other animal faces like we do, and this is all because of the size of our heads, and therefore the size of our brains. Our enormous brains are so valuable that they are, in an evolutionary sense, worth the pain and danger of our childbirth. And yet, even with all that pain, our brain is still nowhere near as developed at birth as that of a chimpanzee or horse. So other mammals' babies can almost immediately partially look after themselves, while baby humans need constant and complete care. If we were born any further along in our brain development, we would have an even higher rate of maternal and infant mortality, and the trade-off might break the other way.

As a result, we need a long development time after birth. For years, our parents look after us very closely, keeping us alive and carrying us around. Many other, particularly large mammals also have long 'childhoods' (elephants, for example, might wean for several years and do not become sexually mature until perhaps 17 years of age). But our helplessness in our first year is unparalleled. Except in birds.

(Some of you might be raising an objection here concerning marsupials—and you are right. Marsupial babies are born even more underdeveloped than humans. However, it is probably best to think of the first months in the marsupial pouch as an extension of pregnancy. For much of the pouch time, the joey is even permanently attached to the nipple, a sort of external cognate to the continuous nutrition from the placenta in placental mammals.[34] As a result, much of the development that takes place in the placental uterus takes place in the marsupial pouch, but the result is the same—by the time the offspring leaves the body of the mother (pouch or uterus), it is much more fully developed than a human baby.)

We are all familiar with the image of a nest of pink, unfeathered, blind, helpless hatchlings. For most birds, life begins with a long period of several months confined to the nest, under the constant care and attention from (usually) two parents. The reason in this case is not head size but egg size. Unlike in the case of placental mammals, a bird's egg yolk needs to contain all of the nutrition the offspring needs to develop—so longer development times need larger eggs with larger yolks. It takes a long time to build a big brain, and so the more brain development that takes place in the egg, the bigger the egg and yolk sac needed. Birds then run up against the same physical problem as humans. Some birds, like the kiwi, do lay enormous eggs for their size, but ultimately, an adult bird can only contribute so much energy to an egg, and can only fit an egg up to a certain size through its body.[35] This places a limit on how much development can happen in an egg, and if the bird in question needs a big, complicated brain, it, like the human baby, needs to be born early in its development, and cared for by parents while it completes its growth.

It is hard to live an r-selected life—living fast, breeding, and dying young—if it takes you most of a year just to leave your parents' constant care. So again, big brains and longevity have good reasons to go together, and we shouldn't be surprised that some of the smartest birds are also the longest-living, like crows or parrots. You might also have noticed that there are some birds, like chickens or ducks, that have hatchlings that are much more self-sufficient—able to walk, feed themselves, and in some cases, even swim within a day or two of hatching. It probably will not surprise you that, for their size, these birds generally live much shorter lifespans than those born helpless (though still longer than mammals), and, while not unintelligent, are not of quite the same cognitive calibre. Here again we see lifespan, intelligence, and level of development at birth or hatching interacting, and here again we see that certain traits tend to urge humans and birds towards an extreme way of life: be born helpless, live a very long time, and be very intelligent. In Chapter 4, on breeding, we will dive deeper into this interaction.

The birds that share our same development pattern—helpless youth, long life, and big brains—are in the superorder Neoaves (Latin for 'new birds'). This group includes all the birds that are not either ratites (like ostriches and emus) and tinamous, or fowl (chickens, ducks, and similar), and contains 95 per cent of all bird species. Neoaves includes everything from wrens to crows to kingfishers to cockatoos, and is one of the most diverse and interesting groups of animals on Earth. It is also one of the smartest, with abounding examples of bird intelligence.

With their long lives, flight, and otherwise similarly extreme lifestyle to ourselves, birds are nature's 'other try' at a really intelligent group of animals. Or perhaps I should say 'first try' because, of course, the birds got there first. The extinction event that killed the dinosaurs and gave birds their flying start also kickstarted them on the road to extreme intelligence, and it took many millions of years for mammals to catch up. That extinction was 66 million years ago, and was followed almost immediately (on geological time scales) by the explosion of the Neoaves on the scene. We did not turn up for tens of millions of years.

* * *

With that kind of a head start, we should not be surprised by some of the impressive feats of intelligence and learning that birds can pull off. Two groups in particular, the parrots and the corvids (crows, ravens, and magpies) have especially developed cognition, and show some of the best examples of avian intelligence (though it can certainly be found in plenty of other groups). Birds, in general, share a lot with primates, in general, in addition to the particularly human traits we have looked at. Crows and parrots, though, might be the closest analogue for our own genus *Homo* in particular. Like us, they have specialized not in particular physical traits, but in a flexible and able mind that can make the best of many different situations.

Parrots have captured the human imagination as intelligent birds for a long time, but crows have been having a particular moment in the past few decades, both in scientific studies and in the public consciousness. Perhaps it is because these less colourful, more shy birds are less attention-grabbing, but crows have been overdue for recognition, and are the subject of much excitement.

Some of the best examples have come from observations by members of the public. Some time between 1970 and 1990, a population of carrion crows in Japan started doing something extraordinary. These crows had discovered that the seeds of a tree planted in an urban park were a good source of food, but the crows were not able to open the seeds themselves. The seeds were in fact walnuts—a great source of nutrients but not one crows can normally eat. The crows had learned through experimentation to place the nuts in the busy road adjacent to the park, and wait for passing traffic to crush the shells, giving them access to the nut inside[36] (Figure 11).

This is already clever behaviour—learning to access a new source of food in a distinctly unnatural environment using the

Fig. 11 Crows like the Japanese carrion crow are assiduous people-watchers, and learn how to live successfully in human cities. This crow has observed the coming and going of traffic at a busy intersection, and learned how the traffic lights predict car movement. She uses this knowledge to her advantage, safely placing hard nuts in the road while traffic stops, so the cars break them open when the lights change.

actions of another species to get the job done. But it is not a unique behaviour—some gulls[37] and eagles have been documented dropping clams or tortoises from great heights to crack their shells in order to access the meat inside, for example. While the Japanese crows' behaviour adds human traffic to the strategy, it is still an analogous use of the surrounding environment to get at a new and rich source of food.

What made the Japanese crows so remarkable was how they learned to navigate in the human city safely. The crows were not simply fluttering out into moving traffic to drop the walnuts. Instead, they would queue up with human pedestrians at a stoplight and crossing, wait for the pedestrian light to signal safe crossing, and walk out into the street to carefully place the walnut on the road in safety. They would then retreat to the pavement while the cars did their work, only venturing out to collect the cracked nut when the traffic light changed again. Not only had they learnt to use the surrounding human activity to their advantage, they had also learnt to observe our traffic signals and connect them to the safety of their activity, building a predictable understanding of their surroundings that allowed them to capitalize on a whole new source of food.

This is the power of intelligence. These crows did not evolve to understand our traffic signals. Rather, over millions of years the crows had developed the ability to learn and to draw multi-layered conclusions about the world around them, and adjust their behaviour accordingly. This is the same sort of ability that has allowed humans to go from living in small bands of hunter-gatherers in African valleys to dominating the planet in gleaming cities in less than 100,000 years—a blip in the evolutionary timescale. Unsurprisingly, crows seem to have discovered this same trick in other parts of the globe as well; since the original Japanese reports, similar behaviour has shown up in California too.

Crows have been casually observed engaging in all sorts of seemingly human behaviour in recent years. In Russia, a hooded crow was observed in 2012 'snowboarding'—the crow had found a plastic disk, probably a discarded lid of some kind, and would repeatedly fly it to the top of a pitched, snow-covered roof, stand atop the disk, and slide down the roof, before flying back to the top.[38] He seemed to be enjoying himself. Crows in Canada have been observed doing the same kind of slides on their backs.

This kind of play can seem remarkably human, and in the case of the Russian crow, a case of tool-use in play. But these, and even the Japanese crows and their walnuts, are anecdotal stories, not scientific studies, and so hard places to draw good conclusions. Many scientists, including corvid expert Alan Kamil, who commented on the above cases when they first made news, are uncomfortable calling the snowboarding crow's behaviour 'play', offering alternative explanations like mating displays or demonstrations of fitness to other crows.

Perhaps the most famous, laboratory-confirmed example of crows' intelligence is the series of experiments performed by my doctoral supervisor and friend, Professor Alex Kacelnik. Alex has had a long career in bird behaviour, including some of the most influential work on risk-taking, foraging strategies, and decision making in a variety of species. His work on crows, however, is his best-known work outside the scientific community, partly, I suspect, because of the usual human fascination with clever animals and with birds. And Alex's crows are seriously clever birds.

Alex worked with a particular species of crow called the New Caledonian crow, found, unsurprisingly, only on the island of New Caledonia. They are not especially large or colourful and look for the most part like a nondescript black crow. But they have one very special characteristic—New Caledonian crows are some of nature's most impressive tool users.

We humans use tools without thinking about it. Our whole civilizations are built on using multiple and compound tools; everything from hammers and levers to computers and automobiles. We make all sorts of devices to help us get jobs done that we could not do with just our bodies. Tool use is a fundamental human characteristic.

It is not at all unique to us, though. Our primate relatives are also great tool users. Chimpanzees and other primates have been studied using rocks as hammer and anvil to smash open nuts,[39] drinking water using a curled leaf as a cup,[40] creating sponges from compressed plant material to soak up water for drinking from hard-to-reach places,[41] and many other impressive tool behaviours. Otters smash open clams with stones.[42] Some dolphins protect their noses with sea sponges while foraging among sharp rocks.[43] Woodpecker finches spear grubs with cactus spines[44] (Figure 12), and there is even a sea urchin that disguises itself by attaching small rocks to its surface.[45] For a long time, humans tried to present tool use as the unique behaviour that defined us and set us apart. Do not fall for it—tool use is too

Fig. 12 Woodpecker finches use sharp cactus spines as tools to skewer grubs to eat, but they haven't been observed using their tools more creatively. Their tool use is an evolved behaviour suited to their environment. It is a great example of well-adapted foraging, but doesn't tell us much about their intelligence.

helpful for other animals not to try it, even animals like the sea urchin that barely have a brain.

Not all tool use is equal though, and the New Caledonian crow is a very special case, not for the fact that they use tools, but for how they use them, make them, and adapt them to new problems. On the surface, the crows use tools in a similar way to the woodpecker finch. The crows use long, thin twigs, or long strips of firm *Pandanus* leaves[46] to extract grubs from deep hollows in trees.[47] The crows hold their tool in the beak, aligned to one side of the head so they can guide the tip with their dominant eye,[48] and thread the tool down the grubs' burrow hole. Then, they poke the head of the grub repeatedly until the grub bites the tool, at which point they pull the grub out of its hole and enjoy a very high calorie, protein-rich meal. It is quite a lot like fishing.

This behaviour alone tends to seem pretty impressive when people first confront it, and leads many to call the crows 'smart'. However, this is an example of the problem of defining intelligence that my biology students run into. This is no doubt an impressive behaviour, but it is a behaviour that the crows have evolved and do fairly automatically. The sea urchins are not very intelligent animals, but have evolved over time to naturally disguise themselves with stones, and the crows' tool use is no different in this respect—it is an inherited behaviour, shaped by evolution. Grubs are one of the richest sources of nutrients on New Caledonia, and the crows have evolved to take advantage of them using tools, because they do not have a big, powerful beak with which to crack through the wood. If they had been parrots, with a strong nutcracker of a beak, they would not use tools, but just chew through the wood, as many parrots are prone to do.

But the crows *are* very intelligent, and we know this thanks to Alex's experiments, and in particular, to a crow named Betty. The experiment that really kicked off our understanding of just how smart these birds are was a simple test of how they choose their tools.[49] In nature, there are lots of long, thin twigs to choose from, some with slight curves, some longer or shorter, and getting the right tool is important to any job. So, Alex and his team had presented the crows with a problem. They placed a morsel of food in a tiny bucket with a curved handle sticking up, and placed the bucket at the bottom of a long tube, held in place with duct tape. Each crow was given a choice of two tools, a straight piece of wire, which obviously could not be used to retrieve the food, and a hooked piece of wire, which could be hooked through the handle to pull up the bucket. This is already a tricky problem and the crows were doing well, showing that they could understand that the hooked tool was the better choice in this test.

Like many great discoveries, things got interesting when something went wrong. Betty and another crow were in the testing chamber working on this problem, when the other crow flew off with the hooked tool and lost it, leaving Betty with no choice but the ineffective straight wire. But instead of giving up or trying and failing with the straight wire, she did something else. Betty jabbed the straight wire into the duct tape to brace it, bent it down into a hook, pulled out the newly hooked tool, and used it correctly to retrieve the food (Figure 13). What's more, she did this time and time again as Alex's team tested and re-tested her to confirm the amazing result. Not only could New Caledonian crows use tools, and choose tools, they could also adapt tools to fit the situation.[50]

Betty's amazing feats kicked off a whole series of experiments to learn just how creative the crows could get. Alex's team tried

Fig. 13 Betty the New Caledonian Crow was supposed to be demonstrating her tool selection skills, choosing either a straight or hooked wire to retrieve a hidden treat. When her aviary-mate flew off with the hook, she shocked the scientific world by bending the straight wire into a hook suitable for the task, one of the first lab-documented instances of a bird making such a tool.

an experiment where the crows had access to a tool that was too short to retrieve a piece of food, but was long enough to retrieve a longer, hidden tool. Sure enough, the crows could use the short tool to get the long tool to get the food, and they could manage even longer sequences of multiple tools just long enough to get at the next. This can be called meta-tool use and the crows are one of the only animals outside of humans who have shown this ability without being explicitly trained; they came up with the solution on their own, and on the spot.[51]

It gets even more impressive: the crows can also build tools. Another of Alex's experiments showed that the crows could combine two short sticks (made to fit together end to end) into one longer tool when the reward was out of reach for the two shorter elements.[52] This is compound tool use, and another ability previously only observed in humans. What makes the crows' behaviour all the more impressive is that, like us, they seem to be able to tell on sight that the meta-tools and compound tools will not work if they skip a step. They do not solve the problems by trial and error, but seemingly by pre-planning. The crows in the meta-tool experiment did not tend to try out the short tool unsuccessfully first, but looked at the problem for a while, and then went straight to the correct, multi-tool solution. The same was true of the compound tools: they seemed to size up the problem and proceed straight to combining sticks to be long enough. Like us, they had the basic physical understanding (sometimes called 'folk physics') to grasp the problem on sight, rather than through repeated mistakes.

What makes all of this so impressive and intelligent, though, is not the tools. Tool use is part of the natural behaviour of New Caledonian crows, so much so that they use tools even when the situation does not require them, like when investigating an unfamiliar object by poking with a stick instead of with their beak. What is impressive is that they can solve problems completely foreign to their natural environment. There are no tiny buckets, or snap-together sticks, or carefully planned meta-tools in New Caledonia. But the crows have general, creative intelligence, and so can solve these problems.

All of this may lead one to suppose that New Caledonian crows are the smartest birds, or perhaps just the smartest corvids—but even that is not true! They are very likely the best tool users of all birds, and so they do very well when tests of intelligence come in the form of tool use. But all corvids do very well in problem-solving tasks, and when those tasks do not centre around tools, the New Caledonian crows still do well, but no more so than any other species of corvid, like ravens or rooks.

Ravens, in particular, have a long history as brainy birds, in our human stories and lore. Aesop cut them both ways. One of his more famous stories, of the Fox and the Raven, has the bird outsmarted by the hungry fox. In Guy Wetmore Carryl's poetic retelling, my favourite by a long stride, the raven is tricked into dropping its morsel of cheese when the fox flatters the bird's singing voice, and asks it: 'pray render with your liquid tongue, a bit from Götterdämmerung?' The bird sings, drops the cheese, and the fox gets its meal.[53]

But Aesop also gave the raven great wits, as in the Raven and the Pitcher. Here a thirsty raven finds a tall pitcher of water, half-full. The raven cannot reach the water from a perch on the pitcher's rim, and so drops pebbles, one by one, into the pitcher, raising the water level bit by bit until he can drink from it.[54] This is a very sophisticated example of folk physics, and a clever bit of problem-solving that many humans don't manage on their first try.

The fable is of course, just that—a story, not science. But it has inspired several researchers to see if their corvids can manage the task. Usually, these experiments have used a food reward rather than just the drink of water. Animals looked after well in high quality labs have plenty of water provided for them, and do not end up parched with thirst like the raven in the fable, so a floating morsel of food is more motivating. But the principle of the experiment is the same. A long tube is half-filled with water, with a floating reward that is out of reach, and a supply of pebbles. This has been tried with both rooks[55] and New Caledonian crows,[56] and both species do manage the trick, dropping in pebbles until the food floats up high enough to reach (Figure 14).

Fig. 14 Aesop told the story of the clever crow who could not reach to drink water from a deep jug. The crow dropped rocks in the jug until the water level rose high enough to drink. Researchers tested New Caledonian crows on the old fable and they solved the problem the same way. They also ignored jugs filled with sand topped with a treat, knowing that the stones would not cause the sand level to rise.

The New Caledonian crows were also given a few twists to make the task a bit harder, like a choice of some pebbles that would sink and some that were Styrofoam and would float (and therefore be useless for raising the reward). They still got it right, and they could even infer that the solution to raising water levels wouldn't work on a similar tube half-filled with sand, which they just avoided. I haven't been able to find an experimental test of ravens in this task, but with rooks and New Caledonian crows so successful, I think Aesop was probably right—the ravens should be able to live up to his story.

* * *

The corvids are a very impressive bunch of animals, and quite exceptional, even among birds. But I would be loath to suggest that other birds do not have truly impressive cognitive skills, and I suspect many of my colleagues, each with their own study species, would be even more adamant on that point. From the learning capacity of songbirds to the complex navigation of migrating seabirds, the diversity of intelligent behaviour among birds is extensive.

Bird intelligence even extends outside the Neoaves, to the more evolutionarily distant and 'primitive' species. Much of my own research has focused on ducks, which, together with geese and swans, make up the Anseriformes, or waterfowl. Together with their landbound cousins, Galliformes, such as chickens and quail, they make up the larger groups commonly called fowl, or Galloanserae, and are the only birds still around that we know to have lived among the (non-bird) dinosaurs. In other words, they are an old group of animals, more evolutionarily primitive, and, it does not hurt me to admit, less intelligent, for the most part, than the Neoaves.

Alex Kacelnik and I originally set up the Oxford Duckling Laboratory to look more closely at a problem we had started researching in pigeons, namely, how much visual information can move from one side of the brain to the other in birds' brains.[57,58] But we soon found ourselves asking deeper questions about what kind of information the ducklings' brains were storing. Unlike the Neoaves, baby ducks do not hatch pink and helpless; as most of us know, they take a few hours to dry off, and then can walk, swim, and follow along behind their mother. One of the very first things a baby duck does is the impressive feat of learning called imprinting. Imprinting is the highly specialized, automatic learning of who your mother is, and knowing to follow her and only her from a very young age. We came across it at the start of this book in relation to Lorentz's experiments. Ducklings are evolutionarily predisposed to do this on the first day of life, long

Plate 1 Cookie the Major Mitchell's Cockatoo holds the avian record for longevity, at least among well-documented examples. As a zoo specimen in the Chicago Zoo, Cookie had good paperwork to attest to his age. Major Mitchell's cockatoos are medium-sized parrots, so it is not unlikely that there are many longer-lived, if less well-documented, macaws and large cockatoos.

Plate 2 The superb fairywren mates for life but is a notorious cheater, both male and female. The offspring of a mated pair may come from a variety of males, but the mated male is not incentivized to leave the pairing, because the young spend a year or more as 'helpers' to rear the next brood. This rearing help offsets the cost of raising some young from a different male.

Plate 3 The superb lyrebird is the most primitive of living songbirds, yet has perhaps the most complex of all songs. It is able to imitate the calls of fellow birds in the Australian bush, as well as human and machine noises.

Plate 4 The Carolina parakeet had the northernmost range of any known parrot, was one of only three species known to have lived as far north as the United States. The species became extinct in 1918, the result of deforestation during the preceding two centuries.

Plate 5 A palm cockatoo holds a drumstick and beats it against a tree hollow. The second largest of all parrots, it is the only one known to naturally use tools in the wild, and the only known animal other than humans to use tools to make music.

before most birds, and certainly humans, are able to learn much of anything.

Imprinting itself is not intelligence. It is a form of learning, and is a good example of an impressive evolutionary adaptation that doesn't involve any innovation, but is an automatic process. But imprinting concealed something more exciting. Alex and I wondered how the ducklings actually conceived of their mother. Imprinting is so quick and automatic that we tend to think of it like taking and storing a mental snapshot of what their mother looks like, and comparing this to objects in the world to see if they match. But it should be pretty obvious that this would not actually work: the mother duck could be at a new angle, or partly behind something, or just stretching her wings in a way the baby hasn't seen before, and she would not match the snapshot.

As it turns out, in addition to learning shape and colour and size, and similar visual qualities through imprinting, the ducklings, only a few hours old, can also learn abstract concepts, like sameness and difference,[59] or heterogeneity.[60] We imprinted ducks on pairs of objects that were either identical or different, and then let them choose from pairs of entirely new objects they had never seen before, but with one pair identical and one different (Figure 15). The ducklings imprinted on sameness preferred to follow sameness, and those on difference followed difference. This seems simple to us humans, but our own children cannot do this for at least a year after birth, and even very intelligent animals like apes and crows need dozens of training sessions to learn to do this. The ducks did it on their first try, with no training, on day one of their lives.

This is not to say the ducklings are smarter than apes or human babies or crows. We took advantage of a naturally quick form

Fig. 15 Despite being only a day old, ducklings can learn relational concepts like 'sameness' and 'difference' in only one training session, and with no reward. This puts them well out ahead of chimpanzees, pigeons, parrots, crows, and even human children, all of whom need many more tries to get it right. The ducklings aren't smarter than these other animals; instead, their rapid, automatic learning, or imprinting, is an important survival adaptation that prevents them from losing their mother in their vulnerable first weeks of life.

of learning in imprinting. Rather, this complex ability to think in analogies seems to be common to all birds. Ducks, being Anseriformes, are evolutionarily older, less intelligent, less specialized than most of the Neoaves. What's more, with similar evidence of analogical thinking in a few Neoave birds, we have good reason to think that from ducks to crows, everything in between can probably think and manipulate information this way. Ducks show us what is common to most or all birds.

* * *

Birds, as a whole, are an intelligent group, and perhaps more consistently intelligent than any other similarly diverse group of animals. Highly intelligent species crop up in a wide variety of animals, across phyla, and across the usual divisions of our own phylum, the vertebrates. Among the invertebrates, the cephalopods, like octopodes and cuttlefish, are very likely the champion thinkers, though honeybees and some spiders are also impressive. The vertebrates have a wide diversity of intelligence, but mammals and birds are where the exciting cognition happens. And I would argue that the birds are perhaps more consistently impressive in this regard. Mammals range from the entirely smooth-brained (if adorable) koala,[61] unable to recognize its only food source, the eucalyptus leaf, when placed on a plate in front of it, all the way up to the primates, including ourselves. Birds do not reach quite the heights of intelligence that humans do, but research suggests that just about every bird species has a quite formidable brain, capable of abstract reasoning. The most intelligent birds, like crows, show a remarkable similarity to our own primate relatives in their abilities. (And I haven't even gotten to parrots yet, but I will.)

So, what ought 'bird-brained' mean? If we are going to continue to throw around that cliché, we may as well make it a bit more accurate and treat it as a compliment. 'Bird-brained' is arguably more likely to indicate a relatively smart animal than 'mammal-brained', and I don't think we'll be brushing that up as an insult any time soon. Again, we see how a life of extremes shared by humans and birds has pushed both groups to a similar way of life, even starting as we did so far apart, as dinosaurs and the rodents between their toes. Long lives and big brains, together not by chance, but because the one amplifies the other. Humans' intelligence evolved in a more extreme way than the birds, who got their extra lift (sorry) from flight, but their flight led them ultimately down the same path. As I said, our big brains got us ultimately to flight, and their flight got them big brains. And it is our big brains and long lives that lead to the next major thing we humans share with birds, and with hardly any other mammals: our family life.

4

TILL DEATH DO US PART

Were you raised by two parents who were not (openly or otherwise) unfaithful? If so, do you know why?

If that is too sensitive, let me ask the same (fundamental) question more delicately: have you ever seen a baby pigeon?

When I first started researching bird behaviour, I was asked by a number of friends and family about one of the little 'folk mysteries' that bemuses children and adults alike. Namely: where are the baby pigeons? Pigeons fill our cities, but have you ever seen a baby pigeon?

It was the sort of question you might get asked by a friend at school or discuss with family over the dinner table while chatting about life's little oddities, and it sticks with me in part because, of course, I *have* seen baby pigeons, hundreds of them! But I understand why most people haven't, or, rather, think they haven't.

Pigeons, or rock doves (the slightly more proper, but still not scientific, name for *Columba livia,* our common friend in cities everywhere), are an adaptive wonder. They are an evolutionary success story in the world of today, affected top to bottom as it is by humans. Rock doves are so called because their original habitat was among cliffs and rock ledges across Europe and at its boundaries with Asia and North Africa. We cannot identify their native range exactly because we have been spreading them around with us for so long.[1] We have domesticated pigeons for eggs, meat, racing, decorative show varieties, and sending messages, and today they can be found in most places where humans are. Many of these populations are *feral*, that is, descended from released domesticated birds, rather than from the wild populations, but in most cases, other than the show varieties, they are essentially indistinguishable from the wild birds. (Though they have been spread so far and interbred so much, it is hard to tell what the 'wild type' might mean any more.[2]) Pigeons are an example of an animal that does extremely well in a world full of humans.

It is not hard to see why. Pigeons are robust birds, plump and sturdy, that can tolerate a wide range of temperature and climates. They can eat just about anything, as tourists in major city centres know all too well. Most importantly, unlike many birds, rather than nesting in trees or burrows or grasslands or any other habitats that we humans tend to destroy, they nest and live in rocky cliffs—the one habitat humans tend to build in vast quantity. We call them 'buildings', but for the pigeons they are just as good, covered with little ledges and cavities perfect for their nests. (At least in the case of traditional architecture—I suspect Bauhaus and Modernism were not well received by the pigeons. Too much glass; too few cornices.) And, thanks to our destruction of all the other types of habitat, there are very few other animals in our cities to compete with or prey upon the pigeons. Humans have filled the world with clusters of food-rich, predator-safe, densely packed pigeon habitats—we just call them cities. And the pigeons have done very well.[3]

For the average city dweller in much of the world, pigeons are probably the single most visible species of vertebrate animal in our lives that is not a pet. In terms of numbers, they are probably beaten by rats or mice in many cases, but the scurrying, nocturnal mammals are much less often spotted than the diurnal pigeons, boldly strutting about our streets and squares as we pass. With such a visible presence, I can see why the seeming absence of babies gives pause for thought to so many people.

When most people think of a baby bird, they conceive of one of two different versions of the idea. On the one hand, there is the barely feathered, waifish, blind hatchling, fallen from a nest in a back garden or park, in mortal danger from cold and starvation, practically unable to move. This is the 'baby bird' that gives rise to the very common but inaccurate idea that it is more harmful than helpful to place the baby back in its nest. Myths abound that the mother will 'reject' her young once contaminated with the smell of humans, or that she pushed the chick out of the nest in the first place. We have in our head as well the more accurate idea that this blind, bald, bulging baby will slowly be fed up by highly attentive parents, becoming fat and fluffy and loudly begging, mouth agape, for another feed, crammed next to siblings in a nest. Let us call this, in Shakespearean style, the 'mewling and puking' model of baby birdship.

On the other hand, we have the other common sight, that of a high-pitched, enthusiastic, ambling, pecking, mischief-making, fluffy ball on legs that can do most of what its mother can, badly. The baby duck, looking for all the world like a duck but fluffy, walking, eating, swimming, and scurrying, or the baby chicken, pecking about the ground at its mother's feet. This model of baby bird still needs warmth and protection, but is an altogether more independent and robust thing than the mewling and puking model. Call this the 'fluffy little adult' version of baby birdship.

Most people, I think, have both of these conceptions in their head when they think about baby birds, and both are indeed accurate. Where I think most people's thinking goes awry is in supposing that birds might go through both of these phases. I can see how someone who has agonized over whether to rescue a helpless baby songbird fallen from a nest, and then fed baby ducks in the park, would come to suppose that: first, birds are mewling and puking; they then develop into fluffy little adults; and finally, they become full-fledged (sorry) adults. They might suppose that fluffy little adult songbirds, given that they grow up in tiny nests high up in trees, are just as much a rarity as seeing the mewling and puking younger babies; similarly, they might assume that hiding somewhere in the reeds at the park is a nest full of mewling and puking baby ducks, undergoing weeks of growth before confidently striding forth as fluffy little adults—though readers will already know otherwise.

This, then, is where the pigeon quandary comes from. Pigeons, though excellent fliers, are birds that spend time out in the open, on the ground, in our cities. Why, then, do we see baby ducks and chickens, fluffy little adults, running about with the adults in the late spring and early summer, but not see the same among the pigeons?

The answer is that this perhaps common conception is wrong. The mewling and puking and fluffy little adult stages are not stages at all, but rather two different types of bird growth, and species do one or the other, not both. Of course, we already know this for the ducks, which, as discussed in Chapter 3, are already walking and swimming, and learning on the first day of life. In fact, these two different models of growing up have scientific names—the

Fig. 16 Bush turkeys are highly nidifugous: their young are hatched highly developed, and largely ready to fend for themselves. Some species can even fly within a few hours of hatching.

mewling and puking version is called *altricial*, while the fluffy little adult version is called *nidifugous*.[4]

The nidifugous birds are the ones whose babies we notice, shadowing their mother or parents, and wandering out in the big world from almost as soon as they hatch. Chickens, ducks, geese, swans, turkeys, and all the similar birds which are in many cases called 'fowl' fall in this category. In the case of Australian bush turkeys (Figure 16), the young are so advanced on hatching that the parents do not linger to provide care at all, and the babies can *fly* as soon as their feathers have dried—a matter of hours.[5]

The altricial birds are the ones whose babies we only see when they have fallen from the nest. These birds spend weeks growing under the careful care and ministrations of the parents, and do not leave the nest until they look very nearly like a fully formed adult. All of the songbirds, crows, parrots, and most other types of

Fig. 17 Pigeons are an altricial species, meaning their babies hatch in an underdeveloped state, and spend an extended period in the nest after hatching under the constant care of their parents. By the time pigeons leave the nest to be seen by humans, they are barely distinguishable from adults.

birds are in this category, as are, as you will by now have guessed, pigeons[6] (Figure 17).

Pigeons seem to us like the sort of bird that we would expect to fall in the first category. They are big (as birds go), roundish like a chicken, spend time on the ground, and are, well, commonplace. So, in our minds we expect them to follow the pattern of a chicken. But, as we learned in Chapter 3, chickens, ducks, and the rest of the nidifugous crowd are Galloanserae, while pigeons are part of the great big Neoaves—more closely related to penguins, petrels, and parrots than to chickens or ducks. The bird groupings aren't completely predictive. Neoaves is a huge, diverse group, and includes birds of both types, from the highly altricial songbirds, to the classically nidifugous grebes. (As a rule, though, birds that are not in Neoaves—including the fowl, and the flightless ratites—are nidifugous.) Pigeons, like other altricial birds, remain in the nest until practically indistinguishable from adults.[7]

If you look carefully, though, you can spot them. They are a bit more slender and slight-looking than the full adults, and the surest sign that you are looking at a baby, only recently emerged from the nest, is the *operculum*. This is the gnarled and desiccated-looking mass of flesh that surrounds a pigeon's nostrils, at the top of its beak, where it meets the rest of the head. If the operculum is still plump, smooth, and glossy-looking, you are looking at a baby pigeon. Often it will look otherwise adult, but still have a tiny, cheeping voice. By the end of the first year, the operculum becomes gnarled, and the bird is grown. 'Baby' pigeons are all around, at least for part of the year—they just already look like their parents by the time we see them.

* * *

Bird classification has always seemed to me to involve repeated cases of saying 'everything except'. When we split birds in two groups, one is small and specific, and the other is 'everything except' those in the first. To some extent, this is always the case for good classifications, but it seems especially pronounced in birds, partly because there are so many species. So, first, we divide all birds (Aves) into Paleognathae, which are the large flightless ratites and the tinamous, and the Neognathae, which are everything except the Paleognathae. Then, the Neognathae are divided into the Galloanserae, the land and waterfowl, and the Neoaves—everything else. The Neoaves get a bit more interesting, with five different subgroups, but even these are not equal partners. Four are quite specific: the hummingbirds, frogmouths, and other big-headed tree dwellers; the rails and cranes; the waterbirds that are not fowl, including everything from flamingos to storks and penguins; and the pigeons and doves. The fifth group, again, is everything else. This continues through several iterations where we peel off the vultures, then the owls, then the falcons, and finally the parrots, leaving the last group of 'everything except', the Passerines.[8]

It might seem, with ostriches, chickens, penguins, parrots, owls, and hummingbirds all comfortably grouped, that the Passerines would be a similarly sized and similarly specific group. But this is why the 'everything except' trend seems so profound for birds. Despite all of the different types of birds we have already excluded, the Passerines comprise over half of all bird species. This makes it a bit difficult to describe the group in the same broad terms as the others, though they are often called songbirds or perching birds. This name suggests that the group would include species like warblers and robins and sparrows, and other prototypical little tweeting birds, and these are certainly Passerines, but the group also includes crows, lyrebirds, bowerbirds, and honey eaters. It is a big, diverse group, but one thing that the Passerines share is that all of them, without exception, are altricial.

Alone, this means that *most* birds are altricial, and despite the famous examples of nidifugous bird behaviour in ducks, geese, and chickens, altricial young are really the overwhelming model in birds. Together with the Passerines, their nearest relatives, the parrots, are also altricial, as are the falcons, owls, pigeons, hummingbirds, and numerous other groups. This means that an overwhelming majority of birds hatch out very immature, helpless young, that require enormous amounts of care over a long period of time.

And if you have ever raised a human baby, you will know from that description that, once again, the birds are behaving a lot like us.

***

Human babies are a fascinating biological phenomenon. Our children are, by a country mile, the most helpless and underdeveloped at time of birth of any placental mammals, and take the longest to develop to the point when they can even partially look after themselves.[9] As we noted before, the marsupials also produce highly underdeveloped joeys, but these young are immediately transferred to the pouch once they are born. There, they latch onto the mother's teat, which swells to fill the mouth, locking the joey in place for months of constant, direct nutrients, after which the young are much more developed and can start to leave the pouch, walking, and feeding themselves. This is a bit like a second pregnancy, and makes for much easier child-rearing than we humans manage.

Among our fellow placental mammals, we are unique. Elephant babies, some of the slowest to grow up and to stop nursing, continue to stay with their mother and nurse for as long as two years or a bit more, longer than some humans. Yet they can walk by two weeks old and start to forage their own supplementary food by three months.[10]

Human childhood is extremely long and extremely dependent by mammal standards, and our young are very, very high maintenance.[11,12,13]

As we learned in Chapter 3, part of the reason we have such a long infancy and childhood is because of our big brains—or rather, our big heads. Humans are rather awkwardly shaped for a live-bearing animal. Our hips are narrow and tilted forward, to a rather extreme degree, when compared to other great apes. Our heads, in turn, are very large compared to our hips, and our heads as infants are especially large compared to our mothers' hips, as well as to our own infant bodies.[14] As a result, our childbirth is extremely dangerous—something we do not share with practically any other vertebrate. Labour takes, at minimum, several hours and can last well over a day, and the actual 'moment' of birth is hardly well described as a moment. A newborn can spend an hour or more in the final push through the birth canal. Chimpanzees, by comparison, birth their whole baby quite quickly, often with a rather climactic moment of birth.[15]

This difficult childbirth is part of the price we pay for our big heads and big brains. Brain development takes a long time, and lots of energy. This development can, in principle, take place both within the womb and after birth, but the downside of long brain development after birth is that the newborn leaves its mother's womb and faces the world with a still very underdeveloped brain. This is the compromise humans face, and that causes us double difficulty. Our babies are born with a huge head that causes painful, dangerous, traumatic birth, and, yet, our eventual brains are so much bigger and require so much more development time that the resulting newborn still requires several years of constant, intensive care before it can even remotely fend for itself. While even the other great apes are born with a brain about 40 per cent of their adult weight, human babies' brains are only 30 per cent of the eventual adult weight, leaving a huge amount of growth to take place outside the womb.[16,17]

The result is that human babies look remarkably altricial. This is not a word that is usually used for mammals, partly because one of the main benefits of mammalian biology is that liveborn young can develop further in the womb than they could if they were in an egg. To us humans, who are all too aware of the discomforts of pregnancy and the pains of childbirth, eggs seem enormously attractive. Lay a reasonably sized egg in (comparative) comfort, look after it for the duration of its incubation (perhaps even with the assistance of an electronic incubator in these appallingly modern times), and following a painless (for the parents) hatching, welcome your new baby into the world. To other mammals, though, whose pregnancies are easier and whose babies have smaller heads and are born straightforwardly, eggs would not seem so attractive. A placental mammal gestating her young in the womb can provide those young with a constant supply of energy and nutrients, limited only by how much she can eat. The mother eats, and the nutrients she obtains can be passed along to the baby through the placenta and umbilical cord. This means that the pregnancy can be indefinitely long, at least from the point of view of providing the young with food. As long as the baby does not get too big to be safely birthed, it can continue to grow safely inside the mother, with her main care duties being to eat more than usual and keep herself safe. Whales have pregnancies well over a year long, and elephants are pregnant for almost two years.[18,19] These massive animals can grow an enormous baby over a long period of time, but one that still can be birthed safely relative to the huge size of the mother, so it makes sense to have a longer pregnancy so as to have a shorter period of vulnerable helplessness after birth.

Eggs, by comparison, are a bad deal. An egg has to contain all the nutrients that a baby needs to develop, right from the time it is laid. This means that, not only can the baby not develop as long, because the amount of nutrient in the egg is fundamentally limited, but instead of spreading out the total developmental nutrition that the baby needs over the course of a pregnancy, the mother needs to provide the whole package of nutrients that the new baby will need all at once. This is a big drain on the mother, and birds can be seriously weakened through the process of laying eggs, because it is so energetically costly.[20,21] Moreover, once the egg has been laid, it is out in the world, separate from the mother, but still, usually, needs to be cared for carefully: protected from predators, warmed and turned, and kept clean. Then, once it hatches, the young must still be raised from their hatching level of development, which in the case of altricial birds, means a very demanding season. If any mothers reading can imagine all the exhaustion they felt during pregnancy condensed into a single egg-laying, which produces an egg that has to be looked after nearly as carefully as a newborn, only to hatch a newborn that still requires that level of care, you can start to understand the downsides of eggs.

Eggs, like babies, have a size limit on what can be safely laid. That size, though, is not merely the size of the final chick which will be hatched, but must also include all the energy that the chick will need through incubation. Some of the nutrients provided in the yolk of the egg will be directly used to construct the body of the growing chick, but that chick will be respiring and burning energy the whole time, so it is not a one-to-one transformation, and a good deal of that mass is lost. As a result, the incubation time for eggs is much shorter than pregnancy time in mammals. The longest egg incubations are in the kiwi,[22] the wandering albatross,[23] and the emperor penguin.[24] The kiwi and albatross both incubate for about 85 days, with interruptions (like foraging), while the penguin sits an uninterrupted 65 days from laying to hatching. This puts a limit on how much development can happen in the egg that simply doesn't exist for mammals. The two bird strategies of altricial and nidifugous reproduction are both compromises on what to prioritize in the limited developmental time allowed by eggs. Some birds prioritize getting the body and basic behaviours up to scratch in time for hatching, and are nidifugous. Others develop the body much less, and instead put the time in the egg towards the first part of a much longer development that continues in the helpless young in the nest—they're altricial. By now, you shouldn't have to guess which type of young tend to develop into the more intelligent, highly developed species of birds: crows, parrots, and songbirds are altricial, while turkeys and chickens are nidifugous.

Because placental mammals do not have the same limits imposed by egg-laying, we don't tend to call their young nidifugous or altricial. If we do try to use those words, though, compared to, say, a baby wren, almost all mammals bear young that are pretty nidifugous: they can do a lot more on day one than the baby wren, and look a lot closer to their adult state, earlier in life. Consequently, making a distinction between nidifugous and altricial mammals is fairly pointless. There is only really one placental mammal that you could call altricial, and it is us.

Humans' big brains are our most important trait, and the curse of our reproduction. They are so valuable that they have caused us to give up many of the benefits that mammalian pregnancy has over egg laying. Every other group of animals, generally, lays eggs. Live-bearing is an evolutionarily rare and highly valuable development, for all the reasons we have already discussed. Mammals are the only group that universally births live young (or nearly, excepting the platypus and the two echidnas), and in other groups it tends to be a rarity—a few species of insects, sharks, snakes, and frogs have also developed similar adaptations. With egg laying being a well-established default in most animals, it takes a large number of complicated evolutionary steps to get to live-bearing. Our heads, and our narrow, upright hips,[25] though, have reimposed some of the limitations of eggs back on humans, and we are stuck compromising on the development of our offspring.

It is common in evolution to have multiple priorities force a compromise, or fall into a self-perpetuating cycle. Our hands are very dextrous and useful partly because we devote a lot of real estate in the brain to their control.[26,27] Our brains are bigger as a result, in order to provide that real estate, and so can direct our hands to do all sorts of clever and interesting tasks. This requires even more dexterity and fine control of our hands, so they need more brain power[28]—a virtuous cycle ensues, promoting bigger brains and better hand dexterity.[29] Then, because our hands and brains work together so closely, we prioritize having our hands free at all times, rather than using them to help us walk and balance.[30,31] This is why we walk upright, rather than hunched over, bracing with our arms, like our nearest relatives, the other apes. Our upright stance, though, causes all sorts of problems, from our comparatively huge and long feet,[32] to our chronic back pain,[33] to our hips too narrow to birth the large brains that need us standing up in the first place.[34] Multiple priorities bounce off each other, accelerate each other, and force us into the position where our biggest advantages, our brains and our upright stance, reimpose the old limits of the egg upon us. Though we don't lay eggs, the practical realities of our dangerous pregnancies and births mean we have to deal with similar time limits, developmental limits, and compromises as the egg-laying Passerines—and, as a result, we have similarly altricial babies. No evolutionary advantage is free, and, once again, humans' unusual approach to mammalian problems has us mimicking bird solutions.

Bearing altricial young does have a number of advantages in addition to being the solution we humans share with Passerines in compromising between brain size and pre-birth development. The Passerines are the group that hosts almost all the cognitively impressive bird species, and this is not unrelated to their altricial young. Altricial young do not have to hatch (or be born) 'finished', their much longer total time in the nest means they can become more complex, particularly in terms of the brain.

In Chapter 3, we looked at a number of different candidate measures of intelligence: total brain size, brain-to-body ratio, and neuron number—all of which have serious problems when it comes to making broad comparisons across species, and which are riddled with exceptions. An alternative predictor of animal intelligence that seems to be more reliable than these other candidates is the relative amount of brain development that happens after birth or hatching, compared to before. Altricial development obviously looks attractive here—the brain starts off less developed, and over a long period of time grows substantially. Nidifugous brains, meanwhile, are born nearly complete, and do not grow too much larger over the course of juvenility and adolescence.

The reason this seems to be predictive of intelligence has to do with our definition of intelligence from Chapter 3—namely, the ability to create new, 'unprogrammed' behaviours based on gathered information and experience. This is why brain mass gained in the womb or egg is less valuable to intelligence than brain mass gained out in the world. The brain mass developed before birth or hatching is—mostly—the pre-programmed brain, grown based on the instructions of the animal's DNA, and loaded with the basic behaviours of life that it will carry out automatically. Brain mass gained after birth is neuron growth and new neural connections that can reflect the learning the animal is undergoing in the world. Thus, having an altricial brain reserves more of an animal's brain development time for when the animal is observing, learning, trialling solutions, and interacting with others—leading to a more intelligent adult brain.[35] Or so the theory goes. Again, comparisons across groups remain difficult. A giraffe is born relatively nidifugous in bird terms, but you would be hard pressed to assert that this makes it less intelligent than an altricial robin. Other factors—here again, the limits on development created by eggs, among other things—still affect the complicated matter of intelligence. What we can say confidently is that within reasonably similar groups of animals, being born with a brain that still has a lot of 'growing room'—that is, an altricial brain—has a big advantage when it comes to adult intelligence over one that hatches out or is birthed fully formed.

The advantages of altricial young, though, come with a big disadvantage. Parenting an altricial baby, as all human parents know well, is an enormous amount of work. Feeding, warming, protecting, cleaning, teaching—every hour of the day can be filled with activity that is critically important to one's precious young. For modern humans, in technological, well-supplied cities, this is exhausting, and at times, frustrating. For animals (and humans), who live much of their lives on the brink of starvation and predation, it is dangerous. A nest full of helpless, demanding young that need feeding more than hourly, and must be kept warm most of the day is a deadly ball-and-chain for a parent. In the case of birds, it robs them of their most advantageous trait—being able to fly away. A mother bird with new chicks cannot flee—even a few minutes away could prove fatal to her young, whether by loss of temperature, starvation, or predation.

It also places enormous energy demands on the parents. For most animals, living is a day-to-day war against starvation. Getting 'enough' is a victory, and abundance is rare. A parenting bird's problem is compounded—the bird has to find enough food to provide for both themselves and their growing offspring, and their own food demands are greater than normal, thanks to all the back and forth trips gathering all that food. Much like new human parents, the parents of altricial birds will spend all day long, morning to night, tending to the survival of their immobile, helpless babies.

Note—'parents', not 'parent'. Rearing an altricial offspring is simply not a one-person—or one-bird—job. An altricial bird species needs at least two adults working to look after the young, for the simple reason that it is impossible to both keep the young warm and feed oneself at the same time. Nidifugous birds have this easier. They do need to sit on their eggs for a few weeks during incubation, but during this time, the only thing the eggs need is warmth and occasional turning. The mother bird (and for nidifugous birds, it is generally, though with a few exceptions, the mother) can leave the eggs for a few minutes at a time to eat and drink, and the eggs retain enough warmth to keep safely developing. She does not need to forage food for a growing baby while away, just feed herself and get back before they get too cool. For some nidifugous birds, like ducks, eggs actually have better hatching rates when their incubation period has periodic cooling. In the duck lab, we set our incubators to simulate the mothers' feeding trips a few times a day, dropping the temperatures for about half an hour to an hour, then bringing them back up again.

Once nidifugous birds hatch, they need warming for another day or so while they dry off, then they are on the move, pecking at this and that, feeding themselves, and following their mother. And in most cases, only their mother. Looking after a nidifugous baby is something that can be done by a single parent, and usually is.

Ducks are one of the best examples of the nidifugous model. With young that are capable of partly looking after themselves, the drake does not have much reason to stick around after mating with the mother. She is the one who lays the eggs, so he is able to mate and leave, and she incubates them, and then protects the young who follow her around as they mature.[36] Some drakes will linger a while to provide a bit of protection during the incubation, but they generally will not participate in rearing the young. This pattern makes evolutionary sense—the male has nothing to gain from helping out: the female is capable of doing the job alone and cannot offload it on him (as he can leave before the ducklings hatch), and he puts himself at risk by sticking around. A duck on its eggs is literally a sitting duck—at the mercy of predators. A mother duck whose nest is found by a hostile predator has two bad options: stay and defend, very likely losing her life in the process, or flee, and leave her delicious, high-nutrient eggs to the predator. This is a huge cost, as each year's brood is important to the duck's reproductive success, and her survival to try again next year is far from guaranteed. Since the female has no choice but to take on the brood rearing, and is capable of going it alone, the male has no incentive to put himself at this same risk. He will defend his mate from other male ducks who want to try to mate with her as well, but once she has laid her eggs, he does not linger.

Despite this, ducks are genteel compared to other single-parenting birds, and many mammals. Ducks mate for the season, and though males may try to compete for the same female, high quality males that are successful in finding a mate will generally stay with her alone throughout the courting and mating phases. Male ducks fathering multiple broods in the same season does happen, but it is not common. This behaviour is called seasonal monogamy and differs from true polygamy because of these yearly commitments to (usually) a single mate. Compare this to another nidifugous bird, the wild turkey, which is truly polygamous.[37] Here, in addition to all the same reasons a father duck has for leaving his mate, we add another: the opportunity of more mates. A male turkey may father multiple broods with multiple females in a season, leaving each female to care for the nidifugous chicks that result. Even if he could help her to protect one or two extra chicks from predators during their vulnerable first year, it would make no evolutionary sense to do so. He can potentially produce dozens more offspring by seeking out more and more mates. Even more so than the drake, he certainly cannot be involved in helping to parent all those broods.

There are exceptions, though, and some of them are among the ducks' closest relatives. Most geese and all swans are species that have biparental care of nidifugous offspring (Figure 18). In these cases, the male will guard the female and the nest throughout the incubation period, will help to protect the young, who imprint on and follow both parents, and continue to provide care and support to the brood and female throughout the process of rearing.[38,39] This may seem surprising because the pressures on male swans are very similar to those of ducks. Just like the duck, the swan puts his life at risk by staying with the female

Fig. 18 Swans mate for life and both parents help to rear the cygnets. Because the adults only need to attract one mate in their lives, the male does not invest in flashy feathers or bold coloration like a mallard. Lifelong mating, biparental care, and similar-looking sexes tend to occur together and reinforce each other.

and standing guard. Why, then, do the swans stay put? The reason is true monogamy. Swans, and many geese, mate for life.[40] They form a pair bond early in life and very rarely 'divorce'. The death of a partner, or a rare divorce, can often be the end of the reproductive life for the remaining partner (though they can and do sometimes re-pair with a new mate). This changes the game for the male swan, as against the male duck: if he were to leave, and leave his mate to the same risk of predation as the duck, he is putting his own future reproductive potential at risk in a way the duck is not. A high-quality duck will, in the normal course of things, find a new mate next year. For a swan, if his mate is killed, so may be his future potential. He has every incentive to stay, defend, and help with the young. If he stays and helps, his mate is also less exhausted by the process, and more likely to still be surviving and thriving for another mating next year. Furthermore, unlike the turkey, he has no other broods to rely on in a given year, so he needs to put lots of effort into making sure that his single brood of cygnets all make it to adulthood.

This is also related to why we can easily tell apart male and female ducks and turkeys, but not swans or geese. Drakes and male turkeys are colourful, with display feathers that show off how healthy and strong they are. These feathers, with their many pigments and iridescent effects, are not 'cheap' to make: they require a lot of energy for the birds' bodies to produce; and they are not 'cheap' to maintain: in the case of the turkey, the male's big tail makes escaping predators more difficult, and in the case of the duck, the male's bright colours make camouflage almost impossible (especially compared to the dappled brown of the females, which can hide easily in vegetation near rivers and ponds). The males' bright colours or long tails are their mating signals to females, saying, 'If I can make this impressive display and keep myself alive while sporting it, you know I am a genetically superior father for your offspring.' The female wants these good genes to contribute to her own offspring so that they will have a better chance of survival, and in the case of her male offspring, a better chance of finding a mate themselves. Seasonally monogamous species, and polygamous species especially, need to invest in these kinds of signals, because they are regularly on the lookout to attract new mates. The duck may have to go out and win a new mate each year, and the turkey has it even harder. A female turkey can only bear one brood per season, and there are about the same number of male and female turkeys. This means that every male turkey that is successful at polygamy, and fathers more than one brood in a season, takes away potential mates from other turkeys.

Many males each year will not manage any broods at all. The males need to maintain their flashy tails and bright red wattles to stand a fighting chance at handing on their genes.[41]

The swans do not have this problem. Once they have found a mate, then, barring catastrophe, they will never need to find another one. It makes no sense to waste resources, or make yourself vulnerable to predators for your whole life in order to sport mating signals that will only be used once. Monogamous species like swans will often attract a mate through rituals instead, like synchronized swimming, head bobbing, singing, or, in the adorable case of mute swans, forming a heart between their gracefully curved necks. As a result, the males and females look more or less identical. This is doubly useful for the males, when you consider that their role of defending the nest during incubation means they need the same ability to hide as their female mate. Bright visual mating signals just don't make sense when you mate for life.

Altricial animals have the same options of polygamy, seasonal monogamy, and true monogamy that the nidifugous ones do. But their more demanding young change the equations, because no matter which way they go, they need to solve the problem that a single parent cannot raise an altricial baby alone.

There are totally polygamous altricial birds, but they are not very common, and calling them polygamous hides the complexities of how they manage to raise their young. Polygamy in the animal kingdom comes in many flavours, many of which involve some sort of group living and social structures that help to spread around the task of rearing young. Cooperation can emerge even among selfish actors because the alternative means no one's young survive. To take an example from mammals, lions can be said to be polygamous—one high-quality, dominant male will lead a pride of multiple females, and will mate with many or all of them. Indeed, the social structure of the pride excludes other adult males to ensure this. Juvenile males may remain in the group with their mother and siblings until they mature, and then strike out on their own to create their own pride. The dominant male is polygamous, but his female mates do not raise their young alone. They raise them cooperatively, with the other females in the pride. Though lion cubs are born quite a bit more developed than altricial birds, as carnivores, they require a fair bit of care. While a newborn deer can soon graze on grasses, a lion must grow up a bit before it can start hunting other large mammals for itself, so it relies more on its mother and its pride. Lions are not altricial, but their prides help them to pull off a kind of polygamy with more success than if each mother went it alone.[42]

This is very different from the true deadbeat-dad polygamy of turkeys. True single parenting like that can only really work with a highly nidifugous species. Altricial birds cannot use the turkey method, but they can use the lion method, or other similar structures, to support kinds of polygamy.

One of the most unfaithful among bird species is the superb fairywren (*Malurus cyaneus*). These wrens have a complicated social life. Like many birds, they mate for life, but this conceals both their many dalliances and their extended households. The fairywren couples stay together for their whole lives, rearing many nestfuls of young, season after season. But while they are socially monogamous, they are notorious for constant and frequent matings outside their pairings, with both male and female copulating with other birds before returning to the pair's nest.[43,44] The result is that any given clutch of eggs may have offspring from multiple males. Ordinarily, this would be a problem for the male—spending time and energy assisting in the rearing of another male's young is a waste of resources, and incentivizes the paired male to avoid contributing to child-rearing. Indeed, in most species, this would be the perfect opportunity for some males to develop a 'cheater' strategy, mating with numerous females and assisting with none of the young, while other males busily help out rearing the cheater bird's young in other nests. This would be a successful strategy at first, but over time its success would be its undoing: the cheating strategy would become so common that there would not be enough non-cheaters to look after all the offspring, and clutches would become less successful.

Fairywrens get around this because, though mated for life in pair bonds, those pairs are not the only birds looking after a given nest. Fairywrens mate for life, but rear children cooperatively, with helpers. A mated pair will live in a group together with up to three helper birds who may or may not be the offspring of one or both of the mated pair.[45] These birds, usually younger than the breeding pair, will assist with defence of the territory and rearing of young. The helpers typically act in this role for a year or two before starting their own group with their own helpers. This takes some of the pressure off the mated couple in the labour-intensive task of rearing altricial young. Enough pressure, in fact, to allow the birds' exceptional promiscuity to persist in an altricial species.

Some readers who fancy themselves the transgressive type (or whose parents did) may perhaps find some elements of the fairywrens' parenting relatable. Look past the promiscuity, and the fairywren model is not unlike the much-touted multigenerational family that shares child-rearing with grandparents, and can also resemble the common practice in large families where older siblings help with new babies and toddlers. But these models, and their counterpart single parents, at the other end of the spectrum, stand in contrast to the commonly accepted and expected default in human child-rearing: two parents who are, at the least socially, and ideally, sexually monogamous.[46] That default is unusual among mammals. Mammalian males mainly leave the females to do the child-rearing alone: after all, the pregnancy and nursing of a placental or marsupial mammal are biologically constrained to be a female job—the male cannot help, no matter how much he might like to. Once the male has copulated and impregnated the female, it is primarily her investment of energy and time that will result in a healthy brood of offspring, and in many mammal species the male leaves entirely after the mating. In other cases, the male remains part of the social group, perhaps as a protector or co-operator in a larger group of animals, and so contributes to the well-being of the offspring inasmuch as he contributes to the well-being and safety of the group. This is certainly true of our nearest relatives, the chimps and bonobos.[47] Both live in groups of multiple males and females, and are generally promiscuous within the groups (the chimp males tend to compete over females and rather jealously guard their mating access, while the bonobos have more of a sexual free-for-all, but both are by no stretch monogamous, even socially[48]). The males continue to be part of the group, but are not directly involved in rearing their own young, who are born, nursed, carried, and protected by their mothers. In forming long-term mated pairs, we humans, even despite the rate of infidelity in our relationships, are unusually monogamous for mammals.

We are much more like birds. We have discussed a number of exceptions, but they are exceptions: most birds are, at least socially, monogamous, either seasonally or for life. And in nearly all altricial birds, that monogamy translates to biparental care of young.[49] Part of this is because the altricial young are very hard to care for, and part of it is because the incentives for males are different. While a male mammal can do comparatively very little to assist in rearing offspring, a male bird can do almost everything the female can. The female bird has one moment of bigger investment right at the beginning: producing and laying a large egg requires vastly more nutrients, energy, and risk than does producing the male's sperm. But that is only the very beginning of the process, and once that egg is laid, the male can start pulling equal weight. In some species, like pigeons, the parents work in shifts, with the male sitting on the egg for the morning and the female in the afternoon, or vice versa. In other species, the female sits on the eggs for the whole incubation, but the male attends to her throughout, feeding and protecting her so that she does not have to leave the nest at all. Once the young hatch, there is no nursing that must be done by the female: both parents can equally fetch food for the young, or sit on the nest to provide warmth. The jobs are more equal, and more demanding than in mammals, and so it makes sense for the fathers to stick around and contribute to the success. As a result, social monogamy (for life or seasonally) together with biparental care form the typical model in altricial birds. Pigeons, parrots, crows, penguins, tits—most passerines and other altricial groups all conform to this model.

The same effect is in place for humans. Our children, like baby birds, require a level of care much higher than other mammals, and that simply outstrips the capabilities of one person. Our altricial young are so demanding that their need of care overcomes the male's ability to leave pregnancy and child-rearing entirely to the female.[50]

One might see a little bit of a contradiction in our monogamy, though. If birds are more monogamous and more prone to biparental care than mammals because of the more equal sharing of rearing tasks, then surely our dangerous, debilitating pregnancies would swing us the other way? Our pregnancies and the danger they pose represent an enormous inequality in contribution to child-rearing, further compounded by the inequality of breastfeeding.

Part of this is simple necessity. If a male human wants to reproduce, he has to ensure the mother of his children survives during the pregnancy, and that the children survive after it. With human pregnancy so challenging and debilitating, the normal mammal approach of leaving the woman to it is simply not an option. A pregnant woman needs help—to do all kinds of things, right down to simply getting around physically near the end of the pregnancy. A pregnant woman alone, without a partner, in our incipient Rift Valley hunter-gatherer days, would have been at a serious survival disadvantage. Furthermore, childbirth and nursing a very young baby can leave a woman exhausted, damaged, and physically fragile. Again, the male has little choice. His reproductive capacity depends on the survival, not just the production, of offspring, and abandoning a brand-new mother to fend for herself is a good way to leave your baby with bad survival odds.

The sheer length of child-rearing in humans also makes our sharing out of work look more birdlike. Human youth is extraordinarily long. We rely on parents to help us make our way in the world for at least a decade. A decade of work to keep us fed, safe, learning, and maturing before we are even remotely able to consider keeping ourselves alive. And only then barely. A 10-year-old can make a sandwich, sure, but 15 might be more accurate for any real chance at successful independence. This makes our pregnancy, for all its enormity, more like the imbalance of the female producing a bird's egg. Yes, it is a huge, unequal element of the reproductive process: but it is just the beginning. Though a female does vastly and enormously more than the male ever could in the first two years of pregnancy and child-rearing, that is perhaps 10–20 per cent of the time it will take to rear that child to independence, and during the rest of that time the male *can* make an equal contribution (even if he doesn't always do so). The relative possible contributions of the male and female humans are a lot nearer to the relative contributions of the male and female altricial birds. Monogamy and biparental care follow naturally, and stand us in contrast to even our nearest mammalian relatives.

* * *

Humans cheat. Quite a lot. So much, in fact, that some argue, in a poorly nuanced and unlettered way,[51] that we are evolved as a promiscuous or polygamous species, and pull together all sorts of salacious evidence to support this, from our uniquely large penises (the largest in both relative and absolute terms of all primates[52]) to the fact that women's breasts are always engorged (most female mammals go conveniently flat-chested when not nursing or about to nurse[53,54]). And to some extent this is true: like ducks and unlike swans, humans have noticeable primary and secondary sexual characteristics that are 'on display' and no doubt reflective of the fact that many members of the species try and succeed to continue attracting additional mates after forming a socially monogamous pairing. However, those arguing that our 'natural state' (whatever that means) is a sort of free-love free-for-all are ignoring the even more convincing evidence of our social monogamy and biparental care, which as we have discussed, are a practical necessity. They are also missing the point that 'cheaters' is a very, very good descriptor of promiscuous humans, especially men. Human sexual infidelity is indeed cheating, both romantically and evolutionarily.

Biologists use the word cheating a lot. Leaving aside its scurrilous meaning for humans, it describes something that commonly happens in the animal world, and can be a powerful component of evolution: the circumvention of the 'rules'—or normal way of doing things—for individual benefit. Of course, the animals do not have rules written down somewhere, so what we typically mean by cheating is a behaviour that hijacks the instinctive behaviour of another animal to work for the cheater's benefit instead of their own. One well-studied type of non-sexual cheating in birds is called brood parasitism. This behaviour is most famous in cuckoos, but is also common to cowbirds and other species,[55] and it is a grand cheat indeed. After mating with a male, a female cuckoo will, rather than laying her egg in her own nest, find the nest of some other species of bird that already has eggs in it, and will sneakily lay her egg together with the others, when the unsuspecting parent is briefly away. Her egg is carefully timed to hatch just earlier than the host's eggs, and her newly hatched chick will usually spend its first moments of life pushing the other eggs, one by one, out of the nest.[56] The unsuspecting parent bird will assume that the cuckoo chick is her own and begin looking after it as she would her own young. The mother cuckoo will often lay her egg in the nest of a much smaller bird, like a warbler, and even

Fig. 19 Cuckoos are evolutionary cheaters. Rather than cheating in their mating pairs, they cheat by taking advantage of other species. A mother cuckoo will lay her egg in another bird's nest, and her chick will monopolize the parenting instincts of the host birds. The baby cuckoo is an expert beggar, and the host birds will keep feeding and caring for him even when he has vastly outgrown the nest and themselves!

her very young baby will become much larger than its surrogate parent before leaving the nest. The baby cuckoo takes advantage of the parenting instinct of the host, displaying its massive, very red mouth, and making rapid begging noises, which together compel the poor bird to keep feeding its monstrous nest-guest[57] (Figure 19). By hijacking the strong parental behaviours of another altricial bird, the mother cuckoo ensures that her offspring is raised with attention and care as a single baby in a well-looked-after nest, and she doesn't have to do any of the work. She has cheated, and very successfully.

Human sexual cheating is a very similar matter, especially for our worst cheaters, men. As we have discussed, human babies, with all their altricial helplessness, take a lot of work to raise, enough work to force us into a biparental, socially monogamous lifestyle. For any given human, it would be very evolutionarily efficient if he or she could produce lots of offspring (and many copies of their genes) without having to go through the labour of raising the children. Most of us, I think, would prefer the rewarding and uplifting task of rearing our own children, but for the entirely amoral evolutionary maximizer, it is a bad deal. Because we do not lay eggs, human females do not have the opportunity to be brood parasites: the baby that a woman births is most definitely (absent expensive medical treatments) her own. But men can be brood parasites. In fact, the similarity between the behaviour of some of men and brood parasitic birds is what has given us the word 'cuckold', from the very same cuckoo who gets some other bird to raise her young. A man who manages to secretly impregnate a woman who is in a stable pairing with a different man is a kind of brood parasite. He produces a child which is his, genetically, but does not participate in that child's raising, leaving that to the other man—the cuckold.

The woman, of course is still raising her own child. So, what is in it for her to cheat? Why risk her relationship with her reliable mate to assist in the cheating man's brood parasitism? One theory is that this allows women to take advantage of the best fathers in rearing their young, while still getting the benefit of the cheater's genes in her offspring. For any female, human or otherwise, mate choice is, in evolutionary terms, an exercise in kitting out her offspring with a set of genes that will make them successful. The female cannot choose which of her own genes each of her offspring will get, but she can choose which male will contribute the other half. The cheating male, as we have seen, has an evolutionary advantage: he can produce lots of offspring for little investment. That is a quality a female might like her sons to have, in evolutionary terms, so that they too can be genetically successful—particularly as that success will also include the half of their genes that she contributed. However, obviously, the cheating male does not make a very good father when it comes to contributing fairly to child-rearing, and helping her look after the babies. If she formed a long-term partnership with the cheater, she would be left pulling most of the weight. Instead, she can choose a long-term partner who is an excellent, attentive, hard-working father to help raise the children, and then go behind his back to mate with the genetically advantageous cheater, so she gets the best of both worlds.[58] The good father, the cuckold, is the victim of both the cheater and his mate.

Modern humans navigating the vicissitudes of love do not, of course, think this way. First and foremost, many humans are now in the surprising and lurid business of preventing their own reproduction—reducing their reproductive fitness. So, the idea that one might strategize to have as many children as they possibly can might seem strange to the modern reader. Furthermore, even for our ancient ancestors, early humans in the Rift Valley, all of this devious calculation was of course, unconscious. No human has ever carefully considered the genetic benefits of cheating—they cheated because they succumbed to emotions. Those emotions, in turn, were carefully tuned over millennia of selection simply because they work, in their environment. The yearning in some humans to be unfaithful to their good and faithful spouse comes from the same selective pressures as the instinct in the cuckoo to lay her egg in another's nest. The desire just happens—the reason is obfuscated. Man is left to decide whether to succumb to evolution's base desires, or resist, and cleave to goodness.

In all cheating, it is critical that most players of the game are not cheaters—if they were, we wouldn't call it cheating, just chaos. This is true of the cuckoo and it is true for human relationships. The cuckoo's brood parasitism only works because most passerine birds are good and attentive parents, with a parenting instinct so strong it overcomes even the oddness of a massive, strange-looking bird in its nest where its own baby ought to be. The more species that become brood parasites, the fewer there are looking after nests for them to parasitize. At some point, the balance tips, and there are so many brood parasites that it becomes a bad strategy: if you are an obligate parasite and cannot find someone else's nest in which to lay your egg, you cannot reproduce. The success of a cuckoo requires that most other birds are not cuckoos.

For humans, this is why our cheating is clandestine, and so taboo. It has to be taboo. If polyamory, free-love, or just open cheating became commonplace and acceptable, men in stable relationships would know that their partner's child is unlikely to be their own, and that they are likely being taken advantage of by both their partner and another cheating man. The incentive to form a long-term pairing (like a marriage) would be eroded, as for men, it would be a recipe for failure. Children would not be raised by two parents but more and more by single mothers, and for such an altricial species as humans, this would be suboptimal for those children, reducing the rate of successful reproduction for the cheating parents. In our pre-modern state, a single-parent child was simply more likely to die. Widespread abandonment of monogamy would ultimately undermine itself. The few humans who did form strictly faithful monogamous pairs would produce healthier, more successful offspring, and so slowly become more common. Over evolutionary time, the equilibrium would be restored, with most humans forming stable, monogamous pairs, and a small cast of cheaters taking advantage of the system just enough not to break it.

Or rather, that is what would happen absent some external force changing the balance. For most modern humans in the developed world, the simple pressures of protection and starvation are usually distant history. We have child care centres, food banks, health care, and numerous other forms of support for humans who are not supported by families. This changes the equation, not so much for cheaters as such, but for single parents. What was once almost impossible is now just very hard. But in that difference lies a serious change to our society. The downside of these good and helpful and compassionate components of modern life is that they raise our carrying capacity for cheaters, for non-committed men, and for the regrettable decline of monogamy.

Most humans have a strong instinctive moral repulsion to cheating, and given our evolutionary history, it is not hard to see why. Our cheating, in fact, is one of the rare things we do not share with birds. Certainly, we cheat, just as some birds do, both sexually and otherwise, but so do many species. Across birds, mammals, fish, reptiles, polygamy abounds and monogamy is rife with cheating. Humans, though, are unique in the repulsion we feel, and our desire to overcome the inclination. We are unique in our moral sense that implores us to remain faithful, and to forego the evolutionary advantages of cheating in favour of the love and care of our partners. We share an enormous amount in our family life and child-rearing with birds, but this difference, this devotion, we should cherish.

5

LEARNING TO SING

Depending on who you ask, either Otto von Bismarck or Benjamin Franklin gave us the overquoted pearl of insight about wise men learning from the mistakes of others, rather than from their own. Which one of them truly coined the idea, or whether it was both, or neither, is immaterial. The point is a good one, and though obvious for us humans, applies broadly across the animal kingdom—or at the very least among vertebrates, and especially birds and mammals. One of the distinct advantages of having a substantial brain and a good capacity for learning is that animals can start to make useful inferences about the world around them, not only from their own lived experience, but also by observing the actions of other animals and their consequences. Usually, other members of the same species.

This ability to learn by watching others rather than doing is variously called observational learning, or social learning, and is one of those behaviours that gets animal behaviourists excited.[1] This is partly because clear demonstrations of animal observational or social learning are not very common, and not always easy to set up experimentally. You need animals that will be able, for example, to watch a conspecific through a window, with interest, but without being overly distracted by the desire to meet, mate, or maim the conspecific as soon as the curtain goes up. Not only that, but you need to select a task for the demonstrator animal to do that is not so easy that the observer would just figure it out on their own under normal circumstances. It also needs to be a task that can be clearly observed visually from the angle of the observer, and it needs to be fairly 'unnatural' so that you can be sure that it isn't something the observer has seen or tried before. These are tall orders when you set out to design an experiment, and there is a real delight for people like myself and my colleagues when a beautifully designed experiment meets all these criteria and shows an animal clearly and straightforwardly learning to do something by watching another animal do it, or at least try to do it.

It is also exciting because successful observational learning provides insights into how an animal thinks. Imagine a bird learning to open a box to retrieve food by watching another bird do it first. In order for the observer to benefit from watching the demonstrator, it needs to draw several cognitive conclusions. It needs to recognize that the other bird is a separate entity from itself. It needs to understand that, despite this, the world works in broadly the same way for the other bird as for itself. It needs to make the connection that the action of opening the box is, in turn, directly responsible for the other bird being able to reach the food. And it needs to realize that the actions taken by the other bird are actions it is capable of taking itself. This is just a sample—there are many more elements the observer has to realize in order to learn by watching. These might seem unimpressive for us humans; we draw all these conclusions automatically, without thinking about them—but all these layers are present in the task of learning by watching. Very likely, all animals that learn this way are also not conscious of all these mental conclusions—like us, they draw them automatically. But their brains need to make these leaps all the same.

Lots of species are capable of observational learning. It is not a trait uniquely shared by humans and birds. Rather, it is one of the most impressive animal behaviours we find across many different groups—if most commonly in birds and mammals. Some of my very first scientific research—before I even really knew what I was doing—was an unpublished experiment into observational learning in the common octopus. Octopodes are well known to be among the most intelligent—if not *the* most intelligent—invertebrates, but they are unusual for being very antisocial animals. They live alone, tend to react violently to their own kind, and only 'socialize' once in their lives, on the single occasion when they mate.[2,3] As we have already seen, behaviours and abilities that are useful in principle do not tend to develop if there isn't strong evolutionary pressure for them, and with no socializing to speak of, octopodes seem especially unlikely to be capable of social learning, for all their intelligence. And yet, despite this evolutionary setback, they can learn from each other. True, they stare through the glass at each other, enraged and highly territorial, but despite their animus, they learn. If you show an observer octopus a fellow octopus opening a puzzle box to retrieve a crab, the observer will usually be able to open the box on his first try, and without much dithering.[4,5]

We see observational or social learning in the wild in chimpanzees, and it seems to form the basis of what can be called 'culture' when it comes to their tool use and culinary habits.[6] Take the example of groups of chimpanzees that have learnt to use pairs of stones as 'hammer and anvil' to crack open difficult-to-eat nuts. This behaviour is localized, and may well have started with a single initial 'inventor' that was watched and copied by other chimps. As of writing, it has spread to several groups, but does not exist in groups of chimps that are physically far away, without regular interaction with the hammer-using groups. The chimps watch each other and learn, and the result is regional culture.[7] A complicated behaviour like hammer-and-anvil use is much easier to learn from a conspecific than it is to invent anew, so we can see the result of social learning by observing how far this new behaviour has spread.

As a result, widespread social learning that disseminates across a population is sometimes (not without controversy) called cultural transmission. The controversy is partly because we tend to think of culture as a particularly human trait, and, indeed, this kind of cultural transmission of learned behaviour is at the very core of what makes humans such an impressive species. We learn from each other: discovering new solutions and sharing them with others (or having them copied). Your reading this book is an act of cultural transmission, as was my writing it. Other scientists have learnt and shared knowledge about birds and brains and apes and octopodes, and I have learnt it from them, and now hand it on to you. But while human culture is doubtless the most complex on Earth, the fundamental elements of what culture is—localized behaviours shared socially—apply across many species of animals.[8,9]

Though these kinds of learning are not unique to birds, birds are masterful social learners. Perhaps the most famous case of cultural transmission in an animal is the case of the widespread robbery of cream from milk bottles across Europe in the 1920s.

In 1921, the town of Swaythling, in southern England, found itself besieged by vandals. Bottles of milk delivered to residents' doors by the milkmen were being found by the residents with their foil caps broken, and the thick, rich cream that gathered at the top of the milk plundered. The residents, bizarrely, at first suspected 'youths' (ah, to think when youthful rebellion was simply stealing cream!). In time, though, the real culprit was caught—the local blue tits (*Cyanistes caeruleus*), small songbirds, were breaking the foil caps and eating the cream soon after the milk was delivered, before the residents opened their doors to collect it. In some cases, the tits were observed waiting for and watching the milkmen, so as to swoop in on the milk as quickly as possible.[10]

This behaviour is clever enough, I suppose, but it did not amuse the people of Swaythling, who found that multiple different attempts to design a tit-proof cap failed. The birds always managed to figure out how to get inside the milk bottles. Over time, though, the problem became bigger than Swaythling. Nearby towns started to report the same problem, and eventually much of southern England was contending with broken milk bottle caps and missing cream.[11] The problem then spread across Europe. The tits were learning from each other, and the knowledge—that these bizarre, unnatural-looking white bottles that turned up in urban areas every morning were full of tasty, high-calorie food—was spread from birds in a single locality across multiple populations covering a continent. It could well have started with a single bird in Swaythling who figured it out, though subsequent research has suggested that there were probably multiple points of origin. As usual, the capacity for flight enabled birds to become an extreme example of a behaviour. Cultural transmission in most mammals is limited by the distance an animal can travel regularly, and of course, by mountains, rivers, oceans, and canyons. For the blue tits, neither the English Channel nor miles of milk-bottle-free countryside stood in the way of cultural transmission. Flight doesn't make cultural transmission happen any easier, but it does make it faster and allows it to spread much further.

Initially, it was suspected that the blue tits were copying each other: watching the actions another bird took to open the milk bottles, and imitating the method. This is an obvious conclusion to draw and is how we often think of human cultural transmission, whether of a song, or a recipe, or hunting method. But imitation is not the only way that cultural transmission can happen. Researchers have found that other species of birds, like North American chickadees, can learn to open a foil-topped bottle quite easily, if placed in an enclosure with one.[12] Blue tits that observe another bird open a milk cap in this way are also not faster at opening their first cap themselves. The method itself is not what is being copied. Instead, it has been suggested that the actual knowledge being transmitted between the actor and the observer is simply the attention paid to the milk bottle. To a blue tit, a sealed glass milk bottle is not very interesting. It gives off no scent of food, looks nothing like food, and might even look menacingly unusual, compared to other more natural-looking objects in the bird's environment. The great innovation of the first milk bottle opener was not so much the method of how to break the foil, but rather, the very fact of flying down to inspect and poke at the milk bottle in the first place.[13] The observing bird, then, sees a fellow blue tit interacting with the bottle without harm, and might even notice that he gets a meal out of his interaction, and so learns to investigate the bottle more closely. The observer can then figure out the relatively simple method of piercing the cap himself.

Researchers call this kind of social learning 'stimulus enhancement'. Unlike imitation, the observer isn't copying the actor's movements exactly, and usually isn't even paying much attention to those actions. Instead, the observer is simply learning that the object being interacted with is safe, of interest to a fellow member of his own species, and may be associated with some sort of benefit, like having food inside. The result of even this simple social learning in a bird like the blue tit is a continent-wide change in behaviour, and the eventual invention of solid metal milk caps.

Learning from each other may well be among the most important human traits, but as we have seen, this kind of learning is not unique to us, but is shared widely across seemingly all types of animals, even the invertebrates. But there is a type of learning from each other so rare and so impressive that it is one of the very few traits that some scientists still argue is completely unique to humans. As usual, I do not think that is entirely true, and perhaps more than any other trait, this is the human ability to which only the birds even come close: language.

* * *

Human language is a marvel. So much of a marvel as to create massive arguments about how it works and how it is learnt. One of the more famous arguments concerns how much of human language is learnt, as opposed to innate. During much of the twentieth century, linguistics was dominated by the idea that some elements of human language, such as grammatical structure, were innate parts of biology, and may even be universal to all humans. We can credit (or blame) Noam Chomsky in particular for developing and promoting this idea, based partly on an argument he called 'the poverty of the stimulus'.[14,15] Chomsky's idea was that human language is so complex, composed of so many words and structures and myriad combinations, that no human was in fact exposed to enough language in their childhood to actually learn it. If you think about the day-to-day life of a child in a family, the child is exposed to the limited number of words and limited kinds of sentences that his or her parents, and other adults, actually use. Despite this, children do consistently learn the correct grammar of their language, rather than a different arbitrary set of wider grammatical rules that could also account for the limited amount of language to which they are exposed. In other words, the total input of language (the 'stimulus') is too limited (the 'poverty') to account for the linguistic abilities a human has after a normal childhood of language learning. Instead, Chomsky argued, there must be some innate sense of grammar that all humans have that makes up the difference.[16]

Intuitively, this feels right. Our language can essentially describe any idea, and uses a very complicated set of grammatical rules to allow it to stack, recombine, and redefine words as needed. Speech to children is simple and straightforward, so it is somewhat surprising that the outcome is the ability to use the full range of human grammar. No one speaks to a child in the future perfect continuous tense—'This time tomorrow I will have been driving for 6 hours already'—but we can all use it. Chomsky's argument was evidently very convincing, because it came to dominate linguistics research for nearly 50 years.

Unfortunately, Chomsky underestimated human learning, and underestimated the quality of the stimulus. After half a century of wasted efforts on innate grammar, more recent research has suggested that the poverty of the stimulus argument is incorrect, and that human language is indeed a learned skill.[17] The complexities of human speech are passed on from one person to another, and it is our incredible learning ability that manages to absorb its rules and vocabulary from minimal input—at least, when we are very young.

Human language probably has no real equal among all animals. But birds, and in particular, the songbirds, come the closest; not only in ability, but also in structure, and in how vocal communication is passed between individuals, and how it develops.

* * *

Lots of animals are capable of *auditory* learning—the ability to learn to recognize sounds and understand what they mean or signal. We are familiar with this from training our dogs, from a herder using specific calls to direct his flocks, or a horseback rider giving commands to his mount. Most land animals communicate with some form of sound among themselves as well, although in many cases these sounds are not learnt but are indeed an innate behaviour of the animal.

What is less common, by far, is *vocal* learning. This is not just the ability to learn to recognize a new sound, but the ability to learn to *produce* that sound. So, while our dogs are auditory learners, and can learn what 'fetch' means, and respond appropriately, they are not vocal learners: our dogs do not turn back to us and say 'no, you fetch!' Vocal learning is a very rare phenomenon but is obviously necessary for an animal that communicates through sounds that are not innate, and it is a kind of social learning. We humans are great vocal learners, especially as children, but even as adults. We spend our whole lives learning new sounds, sayings, and songs, and reproducing (or in the case of English speakers learning French as adults, *nearly* reproducing) those sounds. But it is not a common mammalian ability. So far, it has only been definitively shown in four groups[18] of non-human mammals:

bats,[19] pinnipeds,[20,21] elephants,[22] and cetaceans.[23,24] I suspect these groups of rather unusual mammals are becoming familiar exceptions at this point!

For birds, on the other hand, it is a way of life. There are three groups of birds that have been shown to be vocal learners: the parrots (obviously),[25] the hummingbirds (curiously),[26] and, of course, the songbirds[27]—our old friends, the passerines. Remember, although the ability is confined to only three groups, the passerines alone represent over half of all bird species, so vocal learning is an ability of a healthy majority of all birds and fundamental to the way those birds live. Our human language and the vocal learning of other big, smart, K-selected mammals, such as whales, seals, and elephants, have a lot more in common with bird communication than with most mammal communication, which tends to involve more scents, sign language, and other visual displays than it does sound, and, particularly, complex, learned sound.

The songbirds are such a large group that it can be hard to pin down what they all share. Vocal learning, as one can tell from the very name of the group, is obviously one commonality, but even here, songbirds have a vast variety of different types of vocalization and breadth of learning. The songbirds first emerged in Australia, and this is still where you can hear the oldest and most complex of their 'languages', that of the lyrebirds.[28]

The two species of lyrebird are native to the eastern forests of Australia, and they are some of the largest passerines. Looking at their shape and behaviour, one might at first think they belong with the fowl: like pheasants or peacocks, they are ground-dwelling birds with large feet, round bodies, and, in males, long ornamental tail feathers. It is only when they open their beaks that their true identify as songbirds becomes breathtakingly obvious. Lyrebirds are master mimics and can replicate the complex calls of kookaburras, whipbirds, pilotbirds, and others so well that they can be mistaken for the real thing by the other birds of the Australian bush.[29,30] More than just birds, though, they can replicate all manner of human-produced sounds, from car alarms to camera shutters to chainsaws—though admittedly those sounds were produced by a captive specimen. The lyrebirds perform these displays of mimicry as part of their mating ritual. A male will scrape and clear a performance area and then give a concert of all the sounds he has learnt—the more complex the call, the more attractive he is to potential mates. And though his call is composed chiefly of imitations of the sounds of other birds, he generally does not learn these sounds by imitating those other species, but, starting as a juvenile, imitates the already complex, imitation-rich calls of other lyrebirds. Adult lyrebirds can pick up new sounds direct from the source, but much of the complexity of their calls has been accrued over generations, rather than in a single lifetime. The result is that, while lyrebirds are excellent mimics, their calls have some recognizable patterns in the combination of different imitative sounds; and as I walk in the Australian bush, I can usually tell if I am listening to the lyrebird or the real thing. The lyrebirds have little 'habits' of speech or turns of 'phrase' in common, a shared part of their 'linguistic culture'.[31]

Given the amazing complexity of lyrebirds' songs—by some measures, the most complex among those of all songbirds—one might be surprised to learn that they are the most 'primitive' of all songbirds.[32] It's true! But it is also a concept—a word—that we have to be careful with. 'Primitive' sounds like it means 'simpler' or 'lesser', and there is a tendency in normal speech to use it that way. In the context of evolution, though, it does not mean any of those things, and, as in the case of the lyrebird, it can mean quite the opposite. When we think about evolution and the divergence of species, 'primitive' means, roughly, 'earlier'. More specifically, it means that the species in question diverged earlier from the rest of its group. When songbirds first evolved in Australia, they would have looked and acted much like lyrebirds, and the modern lyrebirds are the descendants of those early songbirds. As the songbirds diversified from one species into thousands, at some early point, some of them, perhaps already fairly similar to lyrebirds, split from the lineages, including the two species of lyrebirds, that remained in Australia. This subset of songbirds that left Australia would go on to become the ancestors of the highly diverse songbirds of the rest of the world. The lyrebird split off from the songbird lineage—'left' the rest of the family—earliest of all. Remaining behind in Australia, they were not subject to the changing environments and selection pressures that their cousins who left experienced as they spread across the world. The result is that, relative to the first songbirds, the lyrebird has changed less than, say, an American robin, which has a lineage that would have traversed at least three continents of different environments in the course of its evolutionary history. That doesn't make the lyrebird any lesser or simpler—it is just an earlier form of songbird, more like the original for its having stayed near where the group first emerged.[33]

Far from the simplicity implied by 'primitive', there is no bird on Earth with a more complex and varied song than the lyrebird. This may seem curious. Why should it be that all the evolution and diversification that came to the passerines after they left Australia have resulted in simpler songs—a winnowing down of abilities, relative to this earliest example?

In a general way, this mirrors the development of human language, at least with respect to the one example of the lyrebird, and the use of sounds in different human languages. Humans, and therefore our languages, originated in Africa and eventually expanded out to other continents. The African languages, though, remain much like the lyrebird's song. When the subset of humans that would go on to become ancestors of all non-Africans left the continent, those who remained behind became the first to diverge from the rest of the human population. We humans became geographically separated much more recently than songbirds, though, and have now reintegrated our whole global population—so there was never time for *Homo sapiens* to form separate species. But we did form different languages, and like the lyrebird, the earliest diverged languages that remained behind on the continent of origin are the most complex—at least with regard to the number of different sounds made.[34]

Linguists use the word 'phoneme' to indicate a specific, discrete sound across languages. It is different from a letter because, of course, sometimes a unique sound is made by multiple letters, and letters do not always make the same sound across languages. So, for example, in both English and Welsh, we have the letter 'u', but in English it indicates a number of different phonemes, from the long sound of 'lute' to the schwa of 'but'. In Welsh, though, it indicates only one, and a different one from English—a long 'e' sound as in 'leek'. (In south Wales, at least. In north Wales it can be a difficult to describe cross between long 'e' and long 'u'—this is the sort of Welsh I learnt, but not the dialect that is usually in learners' books.) Different languages have different numbers of phonemes. English, depending on the version and accent, has about 36–40. French, with its nasal sounds and uvular 'r' that gives learners so much trouble, has a similar number. Welsh reaches up to 48 with, among other additions, its famous and rare 'll' sound—a bit like lisping around the sides of the tongue (I can do it, I just can't describe it very well).[35]

European languages, though, are generally quite light on phonemes. The real lyrebird languages among humans are to be found in sub-Saharan Africa, where languages like Taa and !Kung (or !Xu) can have more than 100 phonemes, higher by far than any other languages.[36] In fact, human languages follow a general pattern when it comes to number of sounds: the further you go from Africa, the fewer phonemes a language will use. As we trace the expansion of the human species across the planet, we also trace a gradual loss of phonemes. It seems that phonemes are easy to lose, but hard to reinvent. This should not be surprising. It is easy to imagine losing one: suppose English lost its 'f', and we substituted an 's' wherever it occurred. There would be a small number of words that would become ambiguous, like 'found' and 'sound', and which over time might change to resolve that ambiguity, but overall, it is an easy thing to accomplish. By contrast, adding a phoneme is hard. Many readers will have tried and struggled to make that French 'r' or Welsh 'll' sound—and those are some of the easier phonemes that English lacks. Adding a new phoneme is a bit like adding a new colour: we can all imagine a world without blue, but we cannot even conceive of a colour that does not exist. Languages seem to work on a society-wide version of this principle—over time, sounds are lost (especially those that are hard to make), and only rarely are they added.

The result is that while African Taa and !Kung have dozens upon dozens of phonemes, Pacific languages far from Africa, colonized by humans only after thousands of years of divergence, have very few phonemes. The same is true of languages in the deep Amazon. The Papuan language Rotokas and the Amazonian Pirahã hold the crown for the smallest number of phonemes, at 11, with Hawaiian trailing behind at 13.[37] The whole Hawaiian alphabet is: A, E, I, O, U, H, K, L, M, N, P, W, and ʻ. (The last is called the ʻokina and is a glottal stop.)

Unfortunately, after the lyrebird, the comparison with bird 'language' falters. Unlike the case of human language, passerine birdsong does not smoothly lose sounds as it radiates away from Australia. There is not such a reliable and clear pattern as there is for humans, but one theory suggests that birdsong gains complexity as it leaves the tropics.[38] Though the earliest songbird does have the most complex calls, for the most part, it is the far northern hemisphere where complex birdsong is generally found. The proposed reason is simple. In the tropics, there is a constant cacophony of sound from the thousands of species that live in these biologically rich areas. Amid all that noise, a songbird wanting to communicate with its own species needs to 'speak' clearly and concisely, with simple, recognizable calls that cut through the racket. Away from the tropics, things are quieter, and songsters like thrushes, wrens, and robins can engage in more varied and complex songs, safe in the knowledge that they won't be drowned out.[39] It is only a general trend, though, and there is a broad geographical distribution of complex song. Still, when it comes to the most complex sounds, both birds and humans started at the top, and left their most complicated sounds on their home continent.

If the history and distribution of human and bird languages only loosely follow the same pattern, where we really find the similarity is in how and when the vocal learning that leads to our complex noise-making happens. Songbirds, parrots, and hummingbirds share with humans the ability to hear and learn new sounds throughout their lives. This is, of course, one of the main attractions of parrots as pets—teaching them to imitate human speech. But we also share, especially with songbirds, a life history that strongly privileges learning our sounds when we are very young. For both humans and passerines, there is a 'critical period' for language learning, and it makes a big difference to our abilities.[40]

Every human who tries to learn a second language as an adult knows that we are quite capable of vocal learning as adults, but that it can be quite difficult. We spend hour upon hour perfecting accents, learning new sounds, words, and sentence structures, and memorizing vocabulary. And we resolve, after an embarrassing linguistic misstep or forgotten word, to raise our children as bilinguals from the beginning, so that they need not struggle in the same way. Babies pick up language remarkably easily (despite the 'poverty of the stimulus') and require no formal teaching to do so—they just listen and experiment with sound-making and rather quickly seem to acquire a whole language. What's more, their 'naturally' learnt language will always be better, more fluent, and more correctly accented than even the most dedicated adult learner's attempts. This is because we humans have a critical window for language learning.[41]

We are born ready to absorb a language, with an amazing ease for acquiring grammar and vocabulary without even noticing. We have this ability because it is essential for our survival: we are a highly social, highly communicative species that lives in a way that requires proficient use of the world's most complicated animal communication. All the collaboration, information-sharing, and interaction that make human life what it is remain closed off to a human without knowledge of the language used. So, a new baby has no choice: our evolution has equipped him or her with a brain just begging to acquire its first language. Then, some time between age 5 and puberty, once a language has been well acquired, the ability fades. This is seemingly because, for an early human, there isn't much selection pressure to learn a second language. All resources, even the flexibility and learning capacity of the brain, are limited, so without a strong need for multiple languages, the ability fades once it has served its purpose.

If you have ever wondered why some languages are more difficult to learn than others, it is partly because of this change in our learning abilities as we age. A captivating account of how human language works can be found in the short book, *What Language Is* by Columbia University professor John McWhorter.[42] I can do but little justice to him, but I'll give my best summary.

Children can pick up on all the little nuances and difficulties of a language without noticing. Difficult tenses, stacking suffixes, highly inflected verbs and nouns: if you learn these in the cradle, they pose no issue; try to learn them in the classroom and they cripple your efforts at fluency. Moreover, it seems that the natural tendency of languages undisturbed by outsiders is to accrue all kinds of complexities, exceptions, and inflections. These can develop from little habits in the normal ways people speak, and so do not have any 'reason' for existing. Many do not add any meaning or detail, but make a sentence decidedly wrong if they are absent. McWhorter calls this the 'ingrown' property of language.

Left to its own devices, complexity accumulates like vines taking over the walls of an abandoned house.

Archi is a minor language spoken by less than one thousand people across seven villages in Russia, just over the Georgian border.[43] It is a rare language, highly localized, and completely irrelevant to international, or even inter-regional interaction. Everyone who speaks it would use another language—like Russian—to speak to anyone from outside its home villages. This means that Archi is only ever learnt by children. There are no second language Archi learners struggling through its grammar as adults. And a good thing too, because they wouldn't stand a chance. Archi is one of the world's most complicated languages. It has a huge number of phonemes, including at least one consonant used in no other known language. More than the sounds, though, the grammar is what sets Archi apart. It is a deeply inflected language, filled with verbs that change endings for everything from tense, mood, gender, aspect, conjugation class, and noun agreement. The result is, astonishingly, that each verb has over 1.5 million possible forms. And most verbs are irregular. Archi is so ingrown and inflected, it is impossible for an adult to pick it up. Only the brain of a child can manage it, so it remains confined to its seven villages. Since the locals need other languages for doing business outside their village, Archi is used mainly for offhand banter, and telling jokes.

Some readers may recall with dread learning Latin. Compared to English, Latin has highly inflected verbs, and one of the difficulties a student faces is learning all those conjugations. But compared to Archi, Latin is easy. Latin verbs have around a hundred or perhaps a few more forms, depending on the verb. Compared to English verbs that barely ever change, that is a lot, but it is within the scope of an adult learner's abilities. Latin is also a very neat language. It lends itself to classroom methods, like drawing up tables of declensions and algorithms of predictable word-parts that the learner simply assembles to create meaning. Need to make a verb future tense? You just insert '-bi-' between the root and the ending. Almost mathematical in the way the parts come together.

Why? Empire. Unlike Archi—very much unlike Archi—Latin has undergone, perhaps more than any other language, centuries of being taught as a second language to many millions of adults. First, as the universal language of the conquering Roman Republic and then Empire; then as the universal language of the Church; then as the universal language of the educated and scientific communities, as well as the Church; and more recently as the universal language of upwardly mobile school children. And the Church. It suits the classroom and the adult learner so much because it has been subject to over 2000 years of classrooms and adult learners. Conquered peoples learning Latin would never have bothered with any of its ancient difficulties—they would just smooth them out and speak a simplified, easy-to-learn version of the language that got the job done. And over time, that became the official version. We see this in English as well. English grammar is very easy. Verb endings, noun cases, gender agreement—all elements of Germanic languages that English used to use—have been polished away by centuries of second-language learners. Languages become easy when learnt by adults, because they have to be.

Songbirds face the same problem. For nearly all passerines, which accounts for most vocally learning species, their language—their song—has to be learnt in youth to be learnt properly. The passerines' vocal learning is, in most cases, constrained in a very similar way to our human language; it has a critical period for learning.[44]

For the lyrebird, vocal learning is an ability that is enjoyed throughout its life. Part of how the lyrebird makes his song appealing to a potential mate is by acquiring and incorporating more and more different sounds to increase the complexity and variety of the performance. The same is true for other birds that use this kind of imitative mating song, like mockingbirds. One might suspect that the songbirds, with their very name attached to their vocal communication, would all share this ability to expand and embellish their songs. Instead, while the other two vocally learning groups, the hummingbirds and parrots, do have lifelong vocal learning and flexibility, the songbirds, minus a few exceptions, have the critical period to contend with.

Songbirds have a prototypical song. The mating song of some species of a songbird is reliable enough that birdwatchers can confidently render the 'lyrics' in guidebooks to allow easy identification of a bird by its song. The American robin sings, 'cheer-up, cheer-a-lee, cheer-e-o'. The canyon wren says, 'tuwee, tuwee, tuwee', slowing down as it goes. The Carolina chickadee says, 'fee-bee-fee-bay', or, according to some 'ca-ro-li-na' (though I suspect that might involve a bit of embellishment on the part of enthusiastic birdwatchers). These are all relatively simple songs, but some have much more complex repertoires. The brown thrasher (*Toxostoma rufum*) can have up to 3000 different songs in its repertoire, interspersing them and mixing them up as they sing.[45] Meanwhile, the zebra finch (*Taeniopygia guttata*), by far the bird with the most studied song, has a few short sounds that are common to all, followed by a complex song that varies between individuals[46] (Figure 20).

Fig. 20 A zebra finch. Research into the zebra finch's song has built the foundation of our understanding of birdsong. Young males listen to their father's song without imitating, then compare their memory of the learnt song to their own attempts at singing, refining their song as they go. The species has a prototypical song, with individual variations in family lines.

In all of these cases, the song is not an innate part of the bird's evolution, but a learned skill, typically acquired by male birds from their father. As with human language, there is a set period in its youth when the young bird is primed and ready to learn these sounds, after which, in many species, new sound acquisition is not possible. (The others, called open-ended learners, can keep learning sounds through their life, but, like humans, they have an easier time of this, and learn faster, during the critical period.[47]) Like humans, these birds learn to communicate in sounds that are understood by the rest of their species or community by imitating the sounds of a parent, but, unlike humans, they do not seem to do their listening and their mimicking simultaneously. With our babies, we tend to talk and teach words bit by bit over the course of years, and human babies start mimicking just a few words first (usually 'mummy', 'daddy' and such), and then keep learning and adding. But for many songbirds, the learning and the mimicking come in discrete steps. First, the young listen to their father and memorize his song. They do not try to reproduce it themselves; they just memorize what it sounded like, and retain that memory. Then they start to mimic. They don't mimic their father's song back to him, though; instead, they compare their own song to the memory they have of their father's song. Like human babies, they start making noises like babbling—trying out noises seemingly at random, testing what sounds their bodies are capable of making. As those sounds start to approximate the memory they retain, they refine and perfect the song until it matches, more or less, the memory of daddy's song.

This may seem a strange way of learning—why not just learn from father right away and start copying him as he sings? The reason is mainly to do with the short time birds have with their fathers, compared to humans, and with the smaller amount of language they are learning. As we have discussed, songbirds have a long infancy, with both parents looking after them. This is a lot of time for child-rearing compared to many animals, but compared to humans, it is still quite brief. While our babies have a number of years with parents to learn language, a baby bird will generally be more or less adult at the end of one year, and has progressively less contact with its parents as he develops, fledges, and leaves the nest. Baby birds typically retain their baby voices quite late into that first year—it is not uncommon to see a more-or-less adult-looking bird still 'cheeping' not too differently from a nestling. This means that the baby typically needs to memorize his father's or his parents' songs before he is physically able to start imitating them, and before their time together diminishes or ends. Happily, even the most complex of repertoires, like the brown thrasher's, is tiny compared to a human language, so it is quite manageable to memorize the whole song before starting to try to match it.

Despite this difference of timing, the whole process, from the critical period to the babbling to the eventual adult fluency, looks so much like human first language learning that it is often used as a model when studying human language.[48,49] This is especially true of birds like the zebra finch that have varying songs. Zebra finches, as I mentioned, have archetypal bits of their songs that are shared across the species, but the complex sections are different across individuals. Within families, though, birds within the same bloodline will share some of those individual song elements. This is not a genetic effect but a social and cultural one. Because each young male learns its song from his father, the idiosyncrasies of the individual songs are handed down the generations, with families sounding like each other.

These variants in the songs of families and bloodlines within a species were, for a time, thought to be a major contributor to speciation in songbirds.[50] The idea here is that these songs are, in one way or another, species identifiers. Their major use is in mating—they communicate to potential mates the quality and desirability of the singer. Part of the reason for the vast diversity of birdsong is so that a given species has a unique enough sound that, in the cacophony of the forest, a female songbird will be able to pick out which singing males are of her species, and turn her attention towards them in making her selection. If you are a chickadee, it is a waste of time going to the effort of singing if you sound just like the robins, and so cannot be found by the other chickadees. Song is also used to claim territory and communicate with conspecifics on all matter of other things, from threats to simply announcing one's presence. All of these uses for song rely on the song being identifiable and unique to a given species.

As families of birds within a species start to sing more and more differently, the theory goes that over time they would eventually drift so far apart in their song that one branch of the species would not recognize or pay attention to the song of another branch. As a result, mating would only happen within groups of similar singers, eventually resulting in two different species. This is a neat explanation for bird speciation that evokes the Tower of Babel and humanity's own mutual unintelligibility, and while there is nothing wrong with the logic of the theory, evidence that this has ever actually happened is mixed. In particular, although different species of birds are distinguished by their different songs, it seems that in just about every case we have studied, it is unclear that the song difference preceded the species separation.[51] So different species do sing different songs, but the latter is not proven to be the cause of the former. And here, human reality delivers a convincing blow to the theory. After all, human languages can be as mutually unintelligible as any birdsongs—and we remain entirely one big, polyglot species.

* * *

Birdsong is not language. Nothing in the animal kingdom comes close to the nuance and complexity, the layered multidimensionality, the flexibility, or the scalability of human language. What we do share with many birds, though, is a system of communication that is learnt, and that relies on a shared understanding of sounds and combinations of sounds that are transmitted from parent to offspring through behaviour, not genetics. Moreover, we share in this, and in many things, the ability to learn from each other. Humans and birds are not unique in this, but are perhaps the best examples of social learning thanks to our learned communication. And it is through birdsong, and the study of its critical periods and learning mechanisms that we have gained some of our insights into how we came to have the complex language that we do.[52]

Yet if birdsong is not language, it is also very much not song—at least not in the human sense. The human penchant for music, for rhythm, is surely unique to us, right? Well, so we thought. But here it seems we share something with the birds as well—or, at least, with one group of birds. At last, we turn to the parrots.

6

PARROTS IN THE MIRROR

At the beginning of this book, I traced the broad story of the parrots, emerging from the ruined world of the dinosaurs, achieving long lifespans and great intelligence, and eventually streaming out of their home continent to dominate much of the Earth. They were, very likely, the most intelligent form of life on the planet for much of the intervening millions of years until humans came on the scene. This is a story remarkably similar to our own. Throughout this book, I have tried to give many examples of how behaviours that we think of as essential and defining aspects of humans are, in fact, common to birds. And we have explored how these behaviours interact and feed off each other to grow more pronounced and more extreme over evolutionary time. We have visited ducks and pigeons and crows and wrens and cuckoos, but I have mostly held back on the parrots. This is not because there are fewer parallels between humans and these colourful birds, nor because they somehow undermine the points I have made in comparing other groups of birds to humans. On the contrary, I have held back the parrots precisely because they are the very best example—our feathered mirror image, and in many senses, nature's 'other try' at a human-like intelligence. They need to be looked at, not as one example of this or that shared trait, but as a cohesive whole; an animal that found, long before we ever did, a parallel evolutionary path to success that would transform humans into the exceptional animal we are today.

Now, I said that 'parrots' and humans are very similar, but, of course, humans, today, are a single species while parrots are a much more diverse group of over 300 species in the order Psittaciformes. There are parrots that are as different from each other as we are from lemurs. It would be a better comparison if I could point to one species of the most human-like parrots and declare 'the sulphur crested cockatoo is the human with feathers'. But I cannot, and this is because of one of the few key differences in the human and parrot histories.

We humans, and our now extinct near relatives, emerged as an exceptional group of species from an otherwise very small family, the great apes. Modern great apes are limited to eight living species in total.[1,2] While they lived, our closest relatives, such as the Neanderthals, made our circle slightly more diverse, but even they eventually went extinct (and we probably had a fair bit to do with their departure).[3] This left a single species, us, to colonize the Earth with great apes, and to do so by reaping the benefits of our bird-like evolution. We have done very well, but we have done so alone, and often in precarious circumstances.

Think back to the impact that killed the non-avian dinosaurs, along with nearly every other large animal. Being large means being rare, because the ecosystem can support fewer individuals, and being large and rare means being at greater risk from random disasters. This may have almost happened to us. There is a theory that suggests a tight bottleneck of human reproduction about 70,000 years ago. At this point, our species would have been around and reproducing for about 230,000 years. However, all modern humans can trace their genes back to a population of between 3000 to 10,000 individuals alive 70,000 years ago.[4] Around that time, a supervolcano in Sumatra erupted, producing what is now Lake Toba, and causing climate change around the globe that may have killed off the vast majority of living humans and other great apes. The proposed extinction would have shrunk our global population down to the size of a small modern town—humanity almost departed history at that point. Our nearest relatives all succumbed to similar risks before they had a chance to hit their stride and become a truly global species, leaving us the only species left in our genus.

The parrots were less susceptible to this problem. Remember why birds live so long? They are, in general, at less risk. Flight, their game-changing adaptation, means they can escape predators and disasters more easily than great, lumbering apes. Add to this that they are smaller, and so less at risk of big-animal rarity (or starvation when food becomes limited), and add further the total randomness of having started their human-like convergent evolution with more species than the apes. Finally, add that they are less violent—at least to each other—than apes can be against other apes, and the result is a less exclusive set of surviving species. The equivalents of our now-extinct near-relatives and 'missing links' are still alive in the parrots.

We should therefore think of 'parrots' as a bit more like 'primates'—a bigger group, with more diversity than a single species or genus. And indeed, Psittaciformes, including all parrots, and Primates, are both phylogenetic orders, so this comparison is about right. Both orders have about 350 species. Which parrots, then, are the 'humans'? I would argue that there are three groups of parrots that exemplify convergent evolution—and parallel history—with humans: the cockatoos, the grey parrots, and the macaws.

These are the really exceptional parrots—the *urparrots*, if you'll allow me. Each one represents the apotheosis of parrottyness on one of the three parrot-infested continents. The large cockatoos rule Australia; the grey parrots lord it over Africa; and the macaws reign in South America. As this distribution suggests, parrots are a southern hemisphere bunch. This leads to another parallel with humans. Both we and the parrots originated in the southern hemisphere tropics (well, we sort of straddled the equator), but we became very much a northern hemisphere species, with about 90 per cent of us found north of the equator throughout history, including today. Parrots went the other way, but with a similarly strong preference, mostly populating the southern hemisphere. The now-extinct Carolina parakeet was so notable precisely for being one of the few northerly parrots, and the only one formerly common in the United States. Modern global travel has finally started to break down the humans to the north, parrots to the south paradigm—there are now almost as many feral ring-necked parrots in London's Kew Gardens as there are feral Englishmen in Sydney's Royal Botanic Gardens.

Many will be familiar with the basic story of human evolution—how we originated in Africa, evolving from a series of increasingly upright, hairless apes. A complex of many factors pushed humans to become as we are—though there is considerable debate on the sequence of things.

For example, our social behaviour, which we share with many birds, and our dextrous and precise hands,[5] which we do not, are both bound up together with our intelligence, as is our diet.[6] Some have suggested that our intelligence, and especially our language, allowed us to become more socially complex,[7] while others argue that our societies got so complex that only the intelligent could manage the mental task of understanding them—pushing us to evolve to become smarter.[8,9] Similarly, walking upright allowed us to become more dextrous with our hands—or did our reliance on hands force us upright?[10] Once we had hands, our brains had to enlarge in order to manage and control our complex hands. Or was it our big brains that allowed us to use our hands more, making them more specialized and forcing us upright? Once we were smart, we invented cooking, which allowed us to access nutrients more efficiently, meaning we could invest more in an even bigger brain.[11] Or was it the other way around?

We know that these qualities interact and reinforce each other—but looking back on human evolution from the present, it is far from clear how this virtuous cycle began and proceeded. What is clear is that a suite of little changes all coincided to render our evolution a bit special and a bit different, driving us to evolve into an extreme sort of ape.

Parrots, particularly the three groups of urparrots, share this trend with us—and this is possibly the most important thing we do share. Like us, they represent the extreme edge of what it means to be a parrot, the case where all the evolutionary forces we have discussed so far combined to push a species or group of species towards an extreme version of its type. What is remarkable, though, is that these two extreme groups of animals, the humans and the urparrots, have, in their extremity, become so remarkably similar. Rather than each being pushed to opposite ends of what it means to be an animal, they have ended up mirror images of each other, with parrots a whispered answer to the question: what if a different group of animals were to evolve like the humans?

It is, of course, more accurate to say that humans are a primate that evolved like a parrot, since the parrots came first. Like the songbirds, they originated in Australia, emerging about 60 million years ago, and at first lacking the big curved beaks and bright colours and brighter behaviour we now associate with the group.[12] We don't know exactly when they first ventured out of the landmass that would become Australia, but it must have happened within their first 10 million years, because we have at least one unambiguous specimen found in Europe within that time.[13,14] Three groups of parrots emerged: the New Zealand parrots, the cockatoos,[15] and everything else (including all parrots found outside Australasia).[16] The New Zealand parrots are the 'time capsule'. They were isolated from all other parrots when New Zealand split off from Australia, and still bear some of the early parrot characteristics, including a thinner and longer beak. Much as the lyrebird gives us a clue about the 'original' songbirds, the New Zealand parrots—keas, kakas, and the kakapo—give us a sort of glimpse of the early parrots. They are similarly unusual as well, especially the kakapo, the largest and the only flightless parrot. Like so many New Zealand species, all the parrots are severely endangered, thanks largely to domestic cats, the scourge of all birds in the antipodes, and everywhere.

The cockatoos and the 'everything else' group—sometimes wrongly called the 'true parrots'—make up the rest of the world, and on each continent that hosts parrots, they produced a small group of species that was, in many ways, natural selection's 'other try' at human-like animals. As a result, for about 30 million years, from the time parrots hit their stride until the great apes appeared, the most intelligent life on the planet was very likely one of these birds.

* * *

There is so much we share with the parrots. It starts, as did our story, with longevity. Recall that the pink (or Major Mitchell's) cockatoo is the record-holding bird, with 83-year-old Cookie taking that prize. This is the longest-lived, reliably recorded bird, but stories abound of large cockatoos and macaws living well past 100. And it isn't just occasional species either. The whole parrot order is well known for its longevity—both an attraction and a difficulty in keeping them as pets.[17] A macaw, African grey, or cockatoo of any description will outlive almost any owner other than a young child, and even smaller parrots like cockatiels, parrotlets, or budgies are a longer commitment than a dog or most cats. Parrots are some of the only pets we keep that must be arranged for in the wills and estates of their owners (a distinction shared with perhaps a few reptiles, like large tortoises—themselves famed for their long lives).

Parrots live for so long for the same reasons as other birds. They fly, and so benefit from the usual strong K selection that affords, and live without many predators, so get a K selection boost there as well. Like humans, though, their great intelligence and adaptability make them wily, and makes living longer, developing a big brain, and investing in long-term survival worthwhile. Even compared to other birds of similar size, parrots live an especially long time.[18]

Parrots also share our monogamy. With the exception of the New Zealand parrots,[19] and a few oddballs like eclectus and vasa parrots, monogamy is the order of the day, with many species mating for life, and the breaking of pair bonds and re-coupling an unusual (if not impossible) occurrence.[20,21] This should not be surprising, given all we have learned. They are long-lived, big-brained species, and, unsurprisingly, altricial. Parrot chicks are extremely delicate and require constant, intensive care, typically involving both parents in some way (whether feeding or guarding, depending on the species). Like us, they face the challenge of raising demanding young with long development times that support great intelligence, a worthwhile investment for their long lives.

Though parrots are perhaps the best and most extreme examples of longevity and monogamy in birds, these are aspects we share with many types of birds. Where the parrots truly become our mirror image is in their brains, where only the corvids come close, and their language—where they stand out from all other animals.

Parrots are famously intelligent, and none more so than the three big urparrot groups. This fame derives in part from the fact that they are able to imitate human speech, though making the sounds themselves is actually not any real evidence of their intelligence. Many birds, from lyrebirds to mockingbirds to hummingbirds, are capable of vocal learning. Not all of them have the throat anatomy that allows them to precisely imitate the sounds of human speech, but the impressive thing about vocal learning is the mental ability, not its physical limitations. At any rate, there are plenty of non-parrots that can be taught to make a few human noises, including some crows. All the same, the fact that parrots are able to imitate our words captivates us, and what they can do with those sounds once learnt is a fascinating aspect of their behaviour—and a touchpoint for our similarity.

Most readers will probably share my own experience of learning about parrots' speech. As a child, I remember learning—or rather, not so much learning as absorbing from fictional and cultural references—that parrots could talk. Or perhaps 'talk'. I do not think I was an unusual child for finding this captivating, and I think the child's fascination with parrots' ability to talk is part of what makes them such popular and well-known birds. There are many colourful tropical birds on this Earth, but the average person—particularly in the temperate Northern Hemisphere—does not know all that much about Gouldian finches or quetzals or birds of paradise, but does know about parrots. All non-domesticated birds (and frankly, even the domesticated ones) are challenging to keep and care for, so I think it is instructive that of the most popular birds kept as pets (as opposed to livestock), all except the finches, pigeons, and the canary, are parrots of one kind or another. Their intelligence and charisma are doubtless part of this, but I think the greatest draw is almost certainly the possibility that they might talk.

Leaving childhood behind is the source of many sadnesses, and for me, one of those was losing the idea that parrots talk to the idea that parrots mimic. At some point, someone—a parent, teacher, or television programme—decided to firmly explain that parrots were capable of copying our sounds but definitely not of understanding them. The parrots that learned to say 'hello' or 'Polly want a cracker', I was told, had learned only that imitating those sounds would get them a reward or food or attention, and that they were just as stupid as all the other birds.

The naysayers were right on one score: parrots are great imitators—like songbirds and hummingbirds, they are vocal learners, and capable of learning and replicating sounds that they hear.

Indeed, compared to the songbirds, at least, they are even more impressive because they can seemingly do so throughout their lives, less constrained by the critical period for song learning in early life.[22] A parrot can certainly learn to simply imitate sounds that it hears, including human speech, without knowing what it means. What I never quite accepted as a child was that this somehow undermined the idea that parrots were intelligent or special for their capacity to talk. Even as a young boy, I thought the ability to learn another species' sounds and consistently imitate them for reward was rather special—and it did occur to me that though I could also learn to imitate sounds I didn't understand (like my Welsh grandmothers' songs), this did not undermine my ability to also learn and imitate sounds that I did understand. Perhaps this was symptomatic of a future scientist, but it seemed like the adults in my life pronouncing the 'science' of parrot speech were engaging in some big non-sequiturs about the parrots' abilities. Maybe I have always been a parrot advocate at heart.

All around the world one may find parrots, both those kept as pets and wild birds that have regular contact with humans, uttering our words. Charlie the macaw (of the dubious Winston Churchill connection) famously had a wide vocabulary of swears and oaths mocking Nazis, obviously taught at some point by a previous, unidentified, velvet boiler-suit wearing owner. In Gerroa, a small beach town a couple of hours south of Sydney, a sulphur-crested cockatoo lives in an aviary outside the fish and chip shop. He is part of a small public collection of animals—a petting zoo, a chicken roost, and a few other parrots—and has picked up 'hello', 'goodbye', and a few other polite phrases that he uses appropriately with passers-by. A few hours up the coast, another of the same species, living wild and spending part of the year near

Balmoral Beach, has learned from countless Australian picnickers to say, less politely, 'piss off!' when approached too closely under his favourite tree. Indeed, recent years have seen increasingly common reports of Australian wild parrots using situation-appropriate expletives, no doubt learning the local variant of English. Most curiously, there seems to be a growing population of wild parrots in Australia that speak a few English words despite minimal contact with humans, who seem to have picked them up from released pets who have re-joined wild flocks.[23] Even if the naysayers were right, and comprehension of meaning was impossible for these birds, it was clear to me that they were doing something impressive: flexibly learning these sounds and at least connecting them to their situational context or the responses they drew from the humans nearby.

One cannot discuss talking parrots, though, without talking about Irene Pepperberg, and Alex, the African grey parrot that was her 30-year experiment.[24] Her work is not without controversy, and has given rise to numerous detractors who try to find every means of accusing Alex of the same lack of comprehension and simple mimicry. But, for me, her work was vindication of a childhood fascination.

Pepperberg started her academic life as a chemist, before becoming fascinated by animal cognition and language, and setting out to study parrot vocalization and test the limits of real comprehension underneath the mimicry. She bought Alex at an ordinary pet store, and particularly chose him for not being an exceptional talker or a particularly interesting bird in his first year of life—she wanted a run-of-the-mill African grey. His name was a contraction of the goal she had in mind, the Avian Language Experiment. She set about teaching him the English words for various shapes, colours, numbers, objects, and a few important words for expressing himself—like 'wanna' for asking for things, 'go' for movement, and 'none', the last one being an especially ambitious concept. Pepperberg trained Alex using the 'model-rival' technique, whereby a human trainer interacts with both the bird and another human, who is asked to complete similar tasks as the bird, and competes for the attention, and rewards, of the trainer.

Alex was slow to learn the words themselves, but his understanding of those words once he had them was remarkable. He could identify most basic colours and shapes, use words like 'key' to correctly identify real metal keys and plastic baby toy keys regardless of how they differed, and knew and could request most of his foods—'banana', 'berry', 'nut', and more by name. He could correctly count objects up to quantities of six,[25] and answer the question 'how many?' accurately, including when it came with a modifier, like 'how many blue?' applied to an array of multiple objects of different colours. He could even use 'none' to correctly answer when none of the objects were the colour asked—an understanding of the concept of zero.[26] Pepperberg and her team would vary the objects, questions, and complications to probe Alex's understanding of the words he was using, presenting him with combinations or problems he had never confronted before in order to demonstrate that he was not simply mimicking or learning by rote. His abilities continued to grow.

These achievements already put Alex in very rare company among animals. The kind of communication Alex was using, with requests like 'wanna nut' and social lubricants like 'I'm sorry' when he got questions wrong or when his humans seemed frustrated, is almost unheard of among animals. We have only witnessed these abilities in ourselves, in Alex (and subsequent African Greys in Pepperberg's lab), and arguably in a few great apes who have learnt to use sign language. Over the 30 years of the experiment, Alex learned well over 100 words, and used them consistently, appropriately, and syntactically. At age 31, he spoke with much of the ability of a human toddler, all the more impressive for his speaking not just a different language from that of his parents but the language of a different species. Alex died unexpectedly one night, a devastating blow to the research programme, given he had another 15 years of average life expectancy at least. He had always been a bit unwell, with respiratory problems and a feather plucking habit, and seemed to have died from a sudden cardiac issue, possibly a genetic timebomb that had been waiting to claim him. Famously, his last (recorded) words were 'You be good. I love you. See you tomorrow', spoken to Pepperberg as she left the lab.

By the time I met and studied with Pepperberg in 2013, Alex had been dead for six years, and work continued with two other African Greys in her lab. Her work had been both much admired and denigrated. I vividly remember an occasion five years later when her work came up in a conference talk, only to have her very name draw a shout from a scientist at the back demanding a private discussion with the presenter to explain why none of her work should be taken seriously. Many of the arguments against her work, I feel, have little behind them beyond the cranky inclination, still prevalent in some corners of the animal cognition world, to doubt all animal achievements out of hand. One of the less apoplectic charges she faces is that she only had one test subject—though given this involved raising a parrot like a toddler for 30 years it is, I think, understandable. In any case, she has moved on to several more subjects since. Having talked with Pepperberg at some length about the experiments, while I understand her critics' concerns, for my part, I am convinced that Alex knew what he was talking about. Pepperberg agrees, calling his vocalizations not 'language', which even she reserves for humans, but 'code'.[27] And a robust code at that. For her, the most convincing linguistic feat was when Alex, confronted for the first time with an apple (for which he had no word), combined his understanding of the colour of berries and the flavour and size of bananas to call the unnamed fruit a 'banerry', the word he continued to use for apples afterwards. This shows a real understanding that the sounds have meaning beyond just the rewards one can eke out of humans, and this ability to embiggen his vocabulary by coining a neologism is among the most cromulent evidence of Alex's understanding of his human speech.

For me, though, Alex's most impressive, and most human achievement was not his last words, or his invention of 'banerry'. One of the games his trainers would play with him was to simply present objects and ask 'what colour?'. Alex was skilled at the game, and could respond correctly and consistently for both single objects, or groups of like-coloured objects among an array of choices. One day, he looked in the mirror at himself and asked 'what colour?', and was answered (by a researcher) with a word he had not learned: 'grey'.[28] He quickly learned the new word, but it was the question that mattered. At that moment, Alex became the first known non-human to ask a question. And it wasn't a one-off fluke. He could apply that and other questions to a variety of situations, often responded to Pepperberg's questions with his own, and wanted them *answered*—displaying frustration when the answers were not forthcoming or were obviously wrong. Amid other talking parrots and birds, and sign-language-using apes, Alex remains the only animal outside our species who has ever asked us a question. There is little an animal could do that would be more like us than to enquire.

***

Like Pepperberg, I still cannot quite call parrot vocalizations 'language'. Our human languages are a unique adaptation that our nearest relatives do not approach, and parrots, our evolutionary mirrors, only rival in shadows and shades. But there is so much more to our mirror-image than language. Parrots share with us intelligence, and in that most human of traits they also share with us learning, fretfulness, personality, and artistry.

Anyone who has undertaken the challenge of keeping an African grey, a macaw, or especially a cockatoo will know how wily they can be. They all make notoriously difficult pets. Macaws are probably the easiest of the three. They are huge, and even more than the others, have enormous beaks that can do some serious damage to an unwanted hand in their cage or near them. Overall, though, they are relatively (and I stress that word) gentle and relatively (and I stress it doubly) quiet. They can draw blood with a bite, or get you evicted for noise, but this can also be avoided through good husbandry, and bird keepers tend to find them the most 'relaxed' of the big parrots. Greys are the smallest of the three, and the least likely to do grievous bodily harm, but the same vocal inclination that makes them some of the best talkers[29] also makes them a noisy housemate, and the same intelligence that makes them a popular bird also drives them to boredom and fretfulness, and they have a tendency to become self-mutilating if stressed or bored, plucking out their own feathers to the detriment of their health and flying ability.[30,31]

A cockatoo is an insane pet to keep. They are gigantic birds, with big, powerful beaks, an inherent, evolutionary drive for ripping things apart, and are perhaps the loudest of all parrots, with a 'call' that sounds more like an adult man screaming in falsetto. Moreover, while some macaw keepers report success at minimizing loud noises by keeping the bird happy and entertained (with the idea that they scream for attention, much like our own children), cockatoos tend to scream in delight—when you get home, when you say hello, when you give them food.

Perhaps this is why, more even than the other two groups of urparrots, the cockatoos, in all their boisterousness, their brashness, their noise, and their larrikin behaviour, are, in my view, the closest that nature has come to causing another truly human-like mind to evolve. They mate for many years or for life, they live for ages, they talk and squawk, and they are wickedly intelligent to the point of being too smart to be brought to heel as a pet (living more like a perpetual toddler in the homes of their 'owners'). Like us, their suite of extreme behaviours feeds on itself—intelligence, social learning and interaction, longevity, and monogamy each push each other and reinforce each other. Like humans as well, their intelligence is almost outsized for the survival problems they face, and spills over into inventiveness, play, and perhaps even artistry. As a result, they are perhaps the most bird-like of birds and the most parrot-like of parrots, and by being so, are also the most human.

One challenge faced by all large parrot keepers, and especially cockatoo keepers, is locking the cage. A bird kept in a cage needs to be actually kept in that cage, and large parrots are well-known escape artists. Unsurprisingly, these geniuses among birds find it trivial to learn to undo the latches, bolts, and clips that keep in small mammals and other birds. Most guides on keeping large birds advise either an actual lock and key, or more convenient but less reliable, a screw-lock carabiner (which poses a difficulty for an animal without hands, at least for a while).

Laboratory research has confirmed the expert lockpicking of these birds as well. The Goffin's cockatoo (*Cacatua goffiniana*) is one of the smaller members of the family and has been the research animal of choice for Alice Auersperg, the sister of Auguste von Bayern, who worked with Alex Kacelnik on the New Caledonian Crows. Noticing the problem-solving ability of these birds, Auersperg, Kacelnik, and von Bayern set them a task that seemed as though it would be impossible. They created a five-step lock that controlled a small hatch, within which was enclosed a cashew—a very desirable prize. When the lock was fully engaged, a cockatoo would have to, in order, pull a pin, rotate and unscrew a screw, lift and remove a bolt, rotate a wheel, and slide a latch to open the hatch and get the nut. The cockatoos were trained to know that the hatch contained the cashew, providing motivation to get it open, but were otherwise unfamiliar with any of the locking devices. It is the sort of physical manipulation problem we spend years watching our children develop the dexterity and understanding to manage.

It took one of the cockatoos 100 minutes to completely solve the lock problem, over the course of five tries of 20 minutes each. Once he had it learnt, there was no stopping him—he never failed again. This bird, Pipin, was the only one to independently learn the ability within the confines of the few sessions offered, but this is where the real human-style intelligence comes in.[32]

The other birds were then given one of two ways to learn more about the locks—some were exposed to incomplete locks with fewer steps between them and the prize, with more steps added one by one, building up their knowledge of the whole lock. The others were allowed to watch Pipin solve the lock. In the end, six of eight birds learned to open the lock, but most of them did this by capitalizing on knowledge already gained, or on the knowledge of others. Critically, once they did learn to solve each step, they rarely ever failed that step again, building up a consistent working knowledge of the lock.

Even more strikingly, the birds also seemed to understand the physical consequences of each part of the lock. The research team finished with trials where a middle section of the lock was missing. This meant not only that the birds did not need to complete that section, but also that all parts of the lock 'earlier' in the sequence than the missing one could be ignored. The birds tended to go straight for the first actually functioning section, indicating that they understood that the other, earlier steps were no longer necessary.

This is an impressive set of behaviours—everything from one of the birds figuring out such a complex, unnatural manipulation problem, to the others learning socially how to do it themselves, to their ability to correctly infer the interaction of the lock parts. It looks positively human. The complexity of the lock rivals or exceeds many early human inventions. But especially remarkable is the understanding—the intellectual familiarity and comfort with the physical interactions of objects in the world—that allows an on-the-fly inference, like the one it takes to skip to the bolt when the screw is missing. That is intelligence facilitating physical laziness—among the most human of all traits.

The Goffin's cockatoos bring that ability for understanding the basic physical interaction of objects around them to the most famous and fraught of animal intelligence exercises as well—good old tool use. Recall that we learned how tool use, once thought to be a defining feature of humanity, is in reality not even a reliable indicator of intelligence, because there are many animals that use tools as an innate part of their behaviour. Goffin's cockatoos, and indeed all cockatoos (with perhaps one exception that we will get to) do not use tools as part of their innate behaviour. Or at least, we think they don't. For an animal as intelligent as a cockatoo, it would not surprise me at all to discover tomorrow that they have been using all sorts of tools for aeons. (Author's note: As it turns out—they might have been! I've left the previous paragraph as it was, in part to illustrate how easily parrots can overcome our long-held assumptions of their abilities. Just as this book was going to press, with the page proofs all finished, a team of researchers, including Alice Auersperg, discovered that some wild Goffin's cockatoos can use a variety of tools to help access seeds in certain fruit. Only a few of the cockatoos do this, and it seems to be an acquired behaviour in response to fruit provided by the researchers, rather than a typical behaviour, but all the same, they are still wild cockatoos using tools. Research is now underway to determine if there is regional variation in tool use—that is, culture—across different populations.)[33]

At any rate, we do not have evidence of universal day-to-day tool use in the species, and it is not a normal part of their behaviour. Our first evidence of tool use by a Goffin's cockatoo came by chance. The same University of Vienna lab that showed us cockatoos' lock-opening skills was home to a male bird named Figaro, who was one of several Goffin's cockatoos living in an aviary. The aviary was built, like most such structures, of wire mesh walls with wood framing that gave it structure, to which the mesh was attached.

A student had been observing Figaro play with a small pebble, which the bird eventually dropped through the wire mesh. The stone fell out of the reach of his beak or claws, try though he might to reach through. So, Figaro found a small stick, and, holding it in his beak, threaded it through the mesh and attempted, unsuccessfully, to nudge the stone back towards him. The researchers were shocked—Figaro had attempted tool use, the first Goffin's cockatoo to be seen doing so. They isolated him from the rest of the flock (to avoid any social learning) and set about testing his tool use more systematically.[34]

They placed a single nut on one of the wood framing beams at the bottom of the aviary, out of reach of his beak or claws, and waited. At first, he found a small stick, unsuited to the task, and quickly discarded it when it became clear that it would not get the job done. Faced with this quandary, Figaro shocked the researchers again. Over the course of 25 minutes he painstakingly chewed off a long, thin piece of the larch wood beam framing, carefully crunching through and cutting the wood to a suitable shape. Once this was detached from the wood, he held it in his beak, threaded it through the mesh, and carefully manipulated the nut towards him until he could snatch it through the mesh and eat it. Not only could Figaro use tools—he could make them (Figure 21).

Figaro's first tool use was already impressive enough—unlike New Caledonian Crows, tool use is not part of the normal, evolved behaviour of a cockatoo, and for good reasons. In the wild, a New Caledonian Crow uses tools to get at food (specifically, grubs)

Fig. 21 Figaro the cockatoo using a tool. Figaro, a Goffin's cockatoo, caught the attention of scientists when he fashioned a reaching tool from a splinter of wood hewn from his aviary's supports. He went on to demonstrate tool use to three other cockatoos, each of whom used a slightly modified version of his method to retrieve a morsel of hard-to-reach food.

hiding within wood logs and branches. Being a crow, it has a long, narrow beak that is great for eating insects and precisely manipulating tools, but cannot break through thick wood. Cockatoos do not have this problem. There is very little in their native Australian forests that their massive, powerful beaks cannot break through. In the wild, foraging cockatoos have been known to do much damage to trees in their pursuit of food. Where a New Caledonian Crow needed to evolve tool use to get food out of tree branches, cockatoos have always had the option to simply chew through the log.

Wire mesh, though, is something for which evolution did not prepare the cockatoo. Faced with the challenge of a material he could not chew through, Figaro deployed his big brain instead of his big beak, and found a stick. If tool use in a regularly tool-using New Caledonian Crow is no sure sign of intelligence, tool use in a cockatoo that has never done so before certainly is. Moreover, if Betty the crow's tool *making* in the hook-bending experiment is a sign of her intelligence, Figaro subsequently making a tool when no suitable one was around is even more impressive still, given that he wasn't using tools at all to begin with.

Figaro demonstrated his intellectual flexibility, and his ability to invent. When Betty bent wire into a hook, she manufactured a type of tool that she had used before. Very impressive indeed. But when Figaro manufactured his wooden stick from the larch wood beam, he invented something never before seen in his species.

That would be enough to draw quite the comparison to human ingenuity. We are, more than anything else, the beast that invents. Our extensive tool-making and modification and use are intimately connected to our intelligence, because it is boundless and always adapting. Figaro, confronted with a problem that his body could not solve, deployed the immediate flexibility of his mind to create a solution with the materials around him—an incredible feat for any species, let alone one that is a two-hundredth of? our weight with a comparatively tiny brain, separated from us by hundreds of millions of years of evolution.

But Figaro wasn't finished. The human way of learning is social, and so is the parrot way of learning—learning from each other is part of what has made both of our species successful. Having isolated Figaro to avoid his passing the gift of tool use on to the other cockatoos in the lab, the researchers then systematically allowed six other cockatoos to observe Figaro using his newfound tool-use skills to retrieve a nut that was otherwise out of reach.[35]

It took only a few such demonstrations to start seeing the invention spread. Interestingly, all three of the male cockatoos who observed Figaro (a male) learned to use tools in a similar way, but none of the females did. This could, of course, be the randomness of a small trial—or one could choose to read the *very* human tendency to sexism into the birds. I think it is more likely that the cockatoos' social learning gives particular attention to behaviours being performed by other birds that are as much like the observer as possible—so a male cockatoo might quite reasonably conclude that the problems faced by other male cockatoos are those most relevant to himself, and give their solutions special attention—a clever and efficient behaviour, if one yet unproven. More important than the sex of the parrots, though, is that it took only four or five observations for half of the test flock to pick up the behaviour—and I would not be surprised if the other three could eventually learn it, given more exposure and practice.

The learning cockatoos also added their own 'spin' to the method they used to deploy the tool—one would manipulate the tool with his foot instead of his beak, while another nudged it along with his tongue. They all learned from Figaro that this sort of tool-using was possible, but further adapted the method to be most comfortable for them. This is no mere copycat behaviour. Much like Alex the African Grey's speech, this is a collection of behaviours that demonstrates real understanding—a grasp of the world and how it works, and an ability to interact with and modify that world in a flexible and ever-changing way. This is a much scaled down, but still recognizable version of what we humans do every day, and of the behaviour that we have used to transform our world.

The great intelligence and resourcefulness that we share with cockatoos arguably also lead them to share some of our nastier traits. The sad truth of humanity is that one of the consequences of our intelligence and inventiveness is our capacity for spite and vindictiveness. This is a very unusual behaviour. Many animals will, unsurprisingly, resort to violence, and even killing, to defend themselves or their offspring or group. Many more still, of course, resort to violence and killing simply to eat. What is most unusual is to resort to violence or destruction to make a point.

Humans do this all the time. All sorts of violence and brutality and unkindness are carried out to send a message—from the enormities of war and capital punishment, with salted earth and heads on pikes, to the little snipes and jabs of office politics, engineered to accomplish nothing more than irritating our fellow humans. This kind of spiteful behaviour—being unpleasant with the intent of sending a message, is a behaviourally complex task, for two reasons. First, both the actor, and the victim, have to be intelligent enough to understand the consequences—and more than that, have to understand that the other party also understands those consequences. So, if I break the handle off your coffee cup because you keep leaving it unwashed in the office sink, I can only rely on the message being effectively sent if I know that you are intelligent enough to connect the two actions together, and possibly be intimidated into changing your behaviour as a result. Otherwise, you might just assume that the breaking was an inexplicable act of nature unrelated to your filthy crockery habits. Furthermore, we both have to understand that the connection between the breaking and the lack of washing is not natural, but occurs only because a fellow human is creating it to make a point. Most animals do not grasp the layered complexity of the spiteful action, and would probably just interpret this as part of the normal level of chaotic unpleasantness that characterizes the natural world.

Second, making a point with spite requires a system of reputation. Otherwise, it is just wasted effort. You would be unlikely to start washing your coffee cup as a result of my behaviour unless you thought it was indicative of how I might continue to behave. If you assumed that I would never do it again, or if you knew I was never visiting your office again, there would be no reason to change your behaviour. It is your expectation that I am a handle-breaking kind of guy that is the ever-looming threat of spiteful destruction that incentivizes your increasingly civilized office behaviour. For most interactions in the animal world, there is not likely to be a reputation effect to carry forward the benefits of spite. Within an animal's consistent social group, reputations can be built and maintained, but a predator and prey animal are unlikely to ever encounter each other again if they both survive their first interaction.

Biologists define spite a little differently from the way we use it in normal language. It is important not to mistake spite for simple violence or destruction, or for careful strategy. A critical element is about benefit. If I snap off your coffee cup handle because I like the handle and I want it, that is not spite. It is unpleasant and destructive, but it is also evolutionarily straightforward. I want something, so I take it, possibly through unpleasant means. Perfectly normal, rational evolutionary behaviour. A mildly famous video of a cockatoo tearing the anti-bird spikes off a building falls into this category.[36] The cockatoo is not being spiteful, just rational; he is removing an unwanted obstacle from his environment. The fact that humans might have liked those spikes to remain in place is irrelevant to him. Real spite has to be something that is of no benefit, or even may be of detriment, to the spiteful actor. For this reason, although we tend to call everything from snapping cup handles to acts of physical violence in war 'spiteful', biologists would not use that term: there is strategy and reputation behind those behaviours that make them rational methods of achieving some benefit. True biological spite accomplishes no personal benefit—it just deals damage to the victim, and sometimes even to the actor.

Real spite is enormously controversial among biologists. Many argue that it isn't really possible, and that any observation we make of seeming spite in an animal (or even in humans) is backed up by some system of reputation we cannot yet observe, or some benefit to the actor that we aren't seeing. Whatever may be the truth, one thing seems at least anecdotally certain: cockatoos can be spiteful.

Outside the realms of laboratory research there is a well-known fact in Australia, shared as a strict warning to human Australians who start to delight at the cockatoos that frequent their back gardens. Many, captivated by the sight of large, boisterous parrots out their back windows, will start to lay out seed and other enticements for the birds, and this is where the warning comes: if you feed them, and later stop feeding them, they will take out their anger on your house. At the time of writing, I am still in the early stages of this process (Figure 22), but I have heard the same advice myself. By reputation, sulphur crested cockatoos in particular will, once you stop, or forget to replenish their seed, rip apart your decking, siding, window trim, electrical cables, garden plants, chair cushions, and anything else in your garden to make clear their displeasure. The New South Wales Department

Fig. 22 The author foolishly feeds sulphur crested cockatoos on his deck in the Blue Mountains, near Sydney. Local lore and New South Wales government documentation warn against feeding cockatoos, who are believed to respond spitefully when more food is not forthcoming, wreaking havoc on decking, window trimming, garden beds, and anything else they can destroy.

of Planning, Industry, and Environment maintains a section of their website delightfully called 'How do I stop a cockatoo from attacking my property?', for this very reason.[37]

Tearing into the wooden trimming is understandable, at first. As we saw in the case of Figaro, pulling apart wood to search for food (though not using it as a tool) is a natural behaviour for cockatoos, and having exhausted the seed in the dish, it might be reasonable for them to conclude that there could be more morsels in your decking, or to briefly thrash a bit of wood framing in frustration. But they continue, meticulously crunching all the way through wood beams well after it is clear they contain no food, and turning to entirely unnatural materials like weather stripping and metal fittings. They wreak devastation on back verandas all across the continent, like mafia men demanding their protection money. Formal research has yet to properly investigate spite in cockatoos, but for at least one nation of humans, it is well known to rival our own.

Though intelligence equips them to imitate some of humans' more monstrous traits, cockatoos also share with us an appreciation for perhaps the most beautiful of all our seemingly unique human inventions.

In 2007, news reports across the English-speaking world concluded with a typical fluffy piece of human interest, or in this case, parrot interest. A cockatoo (again, a sulphur-crested) named Snowball was shown bobbing and moving and swaying and stepping—indeed, dancing—to the music of the Backstreet Boys. The video was remarkable not only for the amusement it gave viewers, but also caught the eye of one scientist, Anniruddh Patel, who noticed something unusual about Snowball's dancing that was not evident in the myriad other videos of animals seemingly moving to music that are posted online every day.

For several years, that was all anybody heard from Snowball. Behind the scenes, though, Patel and his team were in touch with Snowball's carers and arranged to do some testing. Patel played Snowball eleven modified versions of the song he had danced to, slowed or sped up to change the tempo, and filmed the bird's dancing. He also interviewed the carers to ensure that their claim was true: that they had never trained Snowball to make any of these moves—they were all spontaneous responses to music. When Snowball danced for Patel's experiments, he did so without anyone watching or interacting with him—it wasn't for reward, and it wasn't for attention. Snowball danced for Snowball. And the results were surprising.

Snowball had rhythm. His timing wasn't great, and he was slightly (or more than slightly) off-beat about 75 per cent of the time. Yet the 25 per cent of the time he was on beat was important—that was still vastly better than you would expect by chance from an untrained animal, moving at random without respect to the beat of the music. Patel's team declared Snowball the first scientifically confirmed dancing animal.[38]

Research continued, and another study a few years later showed that Snowball had at least 14 different dance moves he used, speeding them up and varying them in different combinations as the mood struck. Stomping, swaying, head-bobbing, and many other rhythmic movements all combined to make unique dances.[39]

Snowball's rhythm, and his untrained rhythmic response to music, made him a very unusual animal. Startlingly few species other than humans have ever been shown to do anything like dancing,[40] and even those that do are controversial. Despite some claims from both sides, it is still unclear if any of the primates are able to dance in the same way as Snowball—creatively, rhythmically, and for their own amusement, rather than reward. Some evidence for the behaviour is emerging in chimpanzees. But in parrots, dancing, if not commonplace, happens often enough to suggest that Snowball was not a one-off or a genius.[41] Several different species have been recorded inventing rhythmic dances, seemingly for their own amusement, without training, and without reward.

This seems partly to do with the same ability for vocal learning that underpins parrots' speech. More than most animals, parrots are attuned not only to the meaning of sounds but also to their production and manipulation, and research suggests that it is this ability that enables their dancing. This seems to be another case of intelligence, and a capacity for learning and invention overspilling its normal evolutionary usefulness and creating behaviours of play and experimentation in a remarkably similar way in humans and in parrots. So far, they are the only other animals that seem to appreciate music in even a shadow of the way we do.

In the far north of Australia's Cape York peninsula, and over the Torres Strait in New Guinea, there is a cockatoo unlike other cockatoos. It is believed to be the evolutionarily oldest cockatoo, diverging from the lineage before all others, and despite being technically part of the 'white cockatoo' subfamily, it is entirely black, save for its bright red cheeks.[42] It is a colossal bird, the second largest of all flying parrots, after the South American hyacinth macaw (*Anodorhynchus hyacinthinus*), with which it shares a disproportionately large beak (even for its size) and a serene, otherworldly look to its eyes and face.

It is the palm cockatoo (*Probosciger aterrimus*) and it is an ancient species that has inhabited these remote regions for millions of years, our closest living connection to the cockatoos that began the long 'out-of-Australia' journey that would eventually bring them into contact with their mirror image that later journeyed out of Africa. Palm cockatoos are enormously long-lived, with at least one scientific report of a zoo specimen reaching 90, and so breed at a leisurely pace, rearing one chick every two years or so. They mate for life.[43] Their communication and vocalization are complex and modular, with a large repertoire of calls and syllables that can be combined and re-ordered,[44,45] and a greeting call that is almost indistinguishable from the human 'hello'.[46]

They are also the only cockatoo we know of that uses tools in the wild. They are adept at selecting and shaping nut-husks and sticks, carefully cutting them with their massive beaks to form the perfect tool. But unlike the New Caledonian Crows, or even their distant cousin Figaro, the Goffin's cockatoo, they do not use the tools to hunt grubs, or reach food placed out of reach by pesky biologists, or to forage at all. They do not, like chimpanzees, use them to smash open nuts, or collect drinking water. The palm cockatoo's tools are not weapons of war or hunting—they are musical instruments.[47] A male palm cockatoo selects a stick or a nut that serves as a drumstick, and beating it against a dead and hollow tree limb, rhythmically drums out a beat that forms part of his elaborate mating ritual as he seeks out his lifelong partner. The drumming can be slow or fast, brief or sustained, and is rhythmic—beating in time as part of the complex task of proving to a female that the male is worth spending a very long life with. Rhythmic drumming is something not even our fellow great apes do.[48] It is perhaps the very core of our human obsession with music, our earliest instrumentation, and shared by all human cultures and peoples. And it is shared with only one animal.

* * *

This book began with an explanation of convergent evolution, and how it provides an insight into the 'why' questions about human behaviour. Though separated by a 300-million-year evolutionary gulf, birds and humans, and, more pointedly, parrots and humans, have converged on several critical elements of behaviour that reinforce and enlarge each other. Parrots began by flying, and humans by becoming radically intelligent. From there flowed long lives, and long childhood development, which gave us monogamous mating and biparental care. Parrots too picked up humans' radical intelligence, and with our long lives and big brains we, and the parrots, started learning from other members of our respective species– sharing knowledge and communicating in increasing detail. Long and highly social lives encouraged the evolution of even more powerful brains, giving way to invention, to play, and, ultimately, to an intelligence so powerful that mere survival is no longer the question, but how to alleviate boredom and engage with the world around us.

Millions of years before us, parrots found this evolutionary groove, and at the time it would have seemed like just one niche among many, one way of living, different from others, but not otherwise extraordinary. It was not—it was a rocket to a life of extremes. When we found the same groove millions of years later, we re-trod and overshot the parrots' path, starting from our own very different mammalian heritage, and ended up looking not like an extraordinary mammal, but like a strangely featherless bird.

In 1787, the first European colony in Australia was established at Sydney, and kicked off the foundation of a new imperial dominion, and eventually a new nation on the continent. The colonization of Australia brought Europeans into contact, and conflict, with one of the oldest human civilizations on the planet. Estimates vary, but Indigenous Australians are thought to have arrived on the continent no later than 50,000 years ago, with some evidence suggesting dates 20,000 years earlier than that.[49,50]

We do not know exactly when the first Indigenous Australians landed on that continent, or exactly how they spread across the land; we can only piece this together from the archaeological sites. Most of what we know about the deep past is limited to what we can dig up from the ground. But we do know one thing that was waiting for them at Cape York, the very north end of the continent, across the vast Gulf of Carpentaria from where they probably first landed in Arnhem Land. For 30 million years, it had been slowly evolving and converging, following the same narrow path through the twists and turns of natural selection to a niche unlike other niches, and now it met its mirror image. It would be tens of thousands of years before biology would reveal how much we shared with our mirror image, but I suspect those first Queenslanders knew. When the first humans arrived in Cape York, there were, of course, no humans there to greet them, but there was a large cockatoo, with a lifelong partner, standing upright, speaking in long, unintelligible but obviously meaningful phrases, looking down on them from the trees. And he was drumming. And he said, 'Hello.'

ACKNOWLEDGEMENTS

The contemplative task of thinking whom one ought to thank in the acknowledgements is compounded in my case by the peculiar sensitivity of this being my first and, so far, only book. The first temptation, given the momentousness one feels as a first-time author, is to thank every soul one can think of, going back to the womb, for nurturing the particularities that led life down the twisty paths to these words on these pages. On the one hand, one is in turn tempted to resist that because, given this is a first book, and not the capstone achievement on a long career of writing (earnestly to be hoped!), that sort of lifetime-achievement-award gratitude-speech litany seems unearned, and unbecoming to the humility of my own position. On the other hand, though, if this is indeed the only book I ever manage to write, I will end up having left very worthy contributors unthanked if I don't thank them now; and if I write many more books, it would be odd to tell the "creation story", writ as gratitude, later on. If I am going to thank everyone, from beginning to end, I have to start at the beginning, and so, litany it is.

That I am in any position to write a book about anything is the work of my parents, Tony and Sandra Martinho. The time and effort they invested—into indulging my curiosity, into my education, and into my relentless questions—made me a scientist, and I am forever grateful to them. My whole extended family nurtured my fascination with the natural world, and I am especially grateful to my aunt and uncle Elena and Mike Wiechmann for the interest they have taken in my education and scientific work throughout my life, and as well to my in-laws, Cathie Hull and Stewart Truswell for their support of my work and family, always.

More than any other one person, I must thank Alex Kacelnik, my doctoral supervisor, research group leader, collaborator, and academic 'father' for . . . everything. Without him, there is nothing—every part of my scientific work has passed under his eyes in one guise or another, and both his own research, and the ways of thinking he taught me, run throughout this book.

I am also grateful to all the members of the Oxford Department of Zoology, whose work and conversations inspired insights, digressions, or anecdotes included here, who supported my research in my time there, and whose friendship enlivened our studies, especially Dora Biro, Tim Guilford, Adrian Thomas, Izzy Watts, Lucy Taylor, Nacho Juarez Martínez, Andrés Ojeda Laguna, and many others.

Beyond Oxford, Auguste von Bayern and Giorgio Vallortigara, among many other researchers, have made a great contribution to my own work and understanding. I add to this a special gratitude for the many undergraduate researchers who undertook projects in the duckling lab, especially the four *Pembrokiensis* who helped to build it.

As this long list of acknowledgements reveals, I cannot write without an editor, and for this I am deeply grateful to Latha Menon at Oxford University Press, who both edited and refined this book, and who first agreed to meet with me to hear the pitch and make the project a reality; and to Jenny Nugee, who oversaw turning a manuscript into a book, and all of the wonderful OUP staff who contributed. Moreover, I am indebted to Sally Davies, my editor at Aeon, who has done a great deal to develop my style and who has helped to promote much of the work that led to this book, and who, most of all, heard and refined this book pitch first, and helped me see sense in abandoning a previous, malformed idea.

There are many more at Oxford I must thank for their support, friendship, mentorship, conversation, inspiration, and input over the years: Jill and Ivor Crewe, Sandy Murray, William Roth, Robin Dunbar, Ralph Walker, Daniel Robinson, Dawn LaValle, Thomas Prince, Jeremias and Abi Adams-Prassl, and countless other colleagues and friends at University and Magdalen Colleges.

At Harvard, my mentors and research supervisors, including Andrew Berry, Naomi Pierce, Ryan Draft, and Florian Engert all set me on the path to studying evolution and animal behaviour, and, in Naples, Graziano Fiorito and Piero Amodio supervised my very first attempts at my own research. It was they who twitched the thread.

My colleagues and friends at St Paul's College, Sydney, have all contributed to making this book possible, and I especially thank the Wardens, Don Markwell and Ed Loane, for the flexibility and accommodation they have made to help my ongoing research fit in with college life and responsibilities, and Katie Allan, first as Associate Dean, then as Senior Tutor, for her constant help in keeping the ship seaworthy. I could not hope for a more scholarly and pleasant community in which to work as Graduate House, and I thank every member past and present for this little islet of academic paradise.

I would also like to thank all of my friends who have shown great interest (and patience) over a number of years as I have developed these ideas, particularly Graham Long, David Barda, Ahir Reddy, Luc Werner, Michael Stanley, and Jakob Lindaas.

Roger and Jane Knight deserve a special thanks for their guidance and encouragement, and for the interest Roger took in the project as a fellow author—the book might not have been finished, and certainly would not have been finished on time, without them.

Finally, and most of all, I am grateful for this and for all things to my wife Emma Martinho-Truswell, for the integral role she plays in everything I do. There is no idea in this book she has not heard and vetted numerous times, no notion that she did not help refine. More than that, I can achieve nothing without her support—she kept me on the straight and narrow through distraction and indecision, and it is as much her book as mine. I am grateful to her and to my daughters, Flora and Clara, for their support in this and in all things.

NOTES

Chapter 1

1. Darwin, C. and Costa, J.T., *The Annotated Origin: A Facsimile of the First Edition of On the Origin of Species* (Cambridge, MA: Harvard University Press, 2009.).
2. von Linné, C., *Systema naturae; sive, Regna tria naturae: systematice proposita per classes, ordines, genera & species* (Lugduni Batavatorum: Haak, 1735).
3. von Linné, C., *A General System of Nature: Through the Three Grand Kingdoms of Animals, Vegetables, and Minerals, Systematically Divided into Their Several Classes, Orders, Genera, Species, and Varieties* (Vol. 4) (Lackington: Allen, and Company, 1806).
4. Rupke, N.A., *Richard Owen: Biology without Darwin* (Chicago: University of Chicago Press, 2009)
5. Heilmann, G., *The Origin of Birds* (New York: Dover Publications, 1927).
6. Qiang, J., Currie, P.J., Norell, M.A., and Shu-An, J., Two feathered dinosaurs from northeastern China. *Nature, 393*(6687) (1998). pp. 753–761.
7. Clarke, J., Feathers before flight. *Science, 340*(6133) (2013), pp. 690–692.
8. Wellnhofer, P., A short history of research on *Archaeopteryx* and its relationship with dinosaurs, *Geological Society London, Special Publications*, 343(1) (2010), pp. 237–250.
9. Grossi, B., Iriarte-Díaz, J., Larach, O., Canals, M., and Vásquez, R.A., Walking like dinosaurs: Chickens with artificial tails provide clues about non-avian theropod locomotion. *PloS One*, 9(2) (2014), p. e88458.
10. Kemp, T.S., *The Origin and Evolution of Mammals* (Oxford: Oxford University Press, 2005).
11. Nilsson, D.E. and Pelger, S., A pessimistic estimate of the time required for an eye to evolve. *Proceedings of the Royal Society of London. Series B: Biological Sciences*, 256(1345) (1994), pp. 53–58.
12. The correct plural, by the way.

13. McGhee, G.R., *Convergent Evolution: Limited Forms Most Beautiful* (Cambridge, MA: MIT Press, 2011).
14. Gatesy, J., Hayashi, C., Cronin, M.A., and Arctander, P., Evidence from milk casein genes that cetaceans are close relatives of hippopotamid artiodactyls. *Molecular Biology and Evolution, 13*(7) (1996), pp. 954–963.
15. Berta, A., Churchill, M., and Boessenecker, R.W., The origin and evolutionary biology of pinnipeds: seals, sea lions, and walruses. *Annual Review of Earth and Planetary Sciences, 46* (2018), pp. 203–228.
16. Jacobs, G.H., Primate color vision: A comparative perspective. *Visual Neuroscience, 25*(5–6) (2008), p. 619.
17. Bennett, A.T. and Théry, M., Avian color vision and coloration: Multidisciplinary evolutionary biology. *The American Naturalist, 169*(S1) (2007), pp. S1–S6.
18. Patterson, F.G. and Cohn, R.H., Language acquisition by a lowland gorilla: Koko's first ten years of vocabulary development. *Word, 41*(2) (1990), pp. 97–143.

Chapter 2

1. United Nations, Department of Economic and Social Affairs, Population Division, *World Population Prospects: The 2015 Revision* (Vol. I) *Comprehensive Tables* (ST/ESA/SER. A/379) (New York: United Nations, 2015).
2. Riley, J.C., Estimates of regional and global life expectancy, 1800–2001. *Population and Development Review, 31*(3) (2005), pp. 537–543.
3. Langner, G., Estimation of infant mortality and life expectancy in the time of the Roman Empire: A methodological examination. *Historische Sozialforschung, 23*(1–2) (1998), p. 299.
4. Scheidel, W., Roman age structure: Evidence and models. *The Journal of Roman Studies, 91* (2001), pp. 1–26.
5. Larson, G. and Fuller, D.Q., The evolution of animal domestication. *Annual Review of Ecology, Evolution, and Systematics, 45* (2014), pp. 115–136.
6. Grimm, D., Why we outlive our pets. *Science, 350*(6265) (2015), p. 1182.
7. Young, R.D., Desjardins, B., McLaughlin, K., Poulain, M., and Perls, T.T., Typologies of extreme longevity myths. *Current Gerontology and Geriatrics Research, 2010* (2010), 423087.
8. Robine, J.M. and Allard, M., Jeanne Calment: Validation of the duration of her life. *Validation of Exceptional Longevity, 6* (1999), pp. 145–161.

9. Coleman, J.T. and Coleman, A.E., A preliminary analysis of Mute Swan biometrics in relation to sex, region and breeding status. *Waterbirds*, 25 (2002), pp. 340–345.
10. Hunter, L., Field Guide to Carnivores of the World (London: Bloomsbury, 2020).
11. Rowley, I. and Chapman, G., The breeding biology, food, social-organization, demography and conservation of the Major Mitchell or pink cockatoo, *Cacatua-Leadbeateri*, on the margin of the Western Australian wheat-belt. *Australian Journal of Zoology*, *39*(2) (1991), pp. 211–261.
12. Mead, J.G., Beaked whales, overview: Ziphiidae. In *Encyclopedia of Marine Mammals* (pp. 94–97) (London: Academic Press, 2009).
13. BBC News. Daughter scotches Churchill parrot claim. (2004) [online]. Available at: http://news.bbc.co.uk/2/hi/uk_news/england/3,417,353.stm [accessed 11 January 2021].
14. Gibbons, A., Did birds sail through the KT extinction with flying colors?. *Science*, *275*(5303) (1997), p. 1068.
15. Meyerowitz, E.M., Plants, animals and the logic of development. *Trends in Cell Biology*, *9*(12) (1999), pp. M65–M68.
16. Fenchel, T. and Thar, R., '*Candidatus* Ovobacter propellens': A large conspicuous prokaryote with an unusual motility behaviour. *FEMS Microbiology Ecology*, *48*(2) (2004), pp. 231–238.
17. Aubert, G. and Lansdorp, P.M., Telomeres and aging. *Physiological Reviews*, *88*(2) (2008), pp. 557–579.
18. Silva, L., Sobrino, I., and Ramos, F., Reproductive biology of the common octopus, *Octopus vulgaris* Cuvier, 1797 (Cephalopoda: Octopodidae) in the Gulf of Cádiz (SW Spain). *Bulletin of Marine Science*, *71*(2) (2002), pp. 837–850.
19. Pianka, E.R., On r- and K-selection. *The American Naturalist*, *104*(940) (1970), pp. 592–597.

Chapter 3

1. Laude, J.R., Pattison, K.F., Rayburn-Reeves, R.M., Michler, D.M., and Zentall, T.R., Who are the real bird brains? Qualitative differences in behavioral flexibility between dogs (*Canis familiaris*) and pigeons (*Columba livia*). *Animal Cognition*, *19*(1) (2016), pp. 163–169.

2. Armstrong, E. and Bergeron, R., Relative brain size and metabolism in birds. *Brain, Behavior and Evolution*, 26(3–4) (1985), pp. 141–153.
3. Jerison, H.J., The evolution of diversity in brain size. In *Development and Evolution of Brain Size: Behavioral Implications* (pp. 29–57) (London: Academic Press, 1979).
4. Kojima, T., On the brain of the sperm whale (*Physeter catodon* L.). *Science Reports of the Whales Research Institute Tokyo*, 6 (1951), pp. 49–72.
5. Ridgway, S.H., Carlin, K.P., Van Alstyne, K.R., Hanson, A.C., and Tarpley, R.J., Comparison of dolphins' body and brain measurements with four other groups of cetaceans reveals great diversity. *Brain, Behavior and Evolution*, 88(3–4) (2016), pp. 235–257.
6. Marino, L., Brain size evolution. In *Encyclopedia of Marine Mammals* (pp. 149–152) (London: Academic Press, 2009).
7. Herculano-Houzel, S., The remarkable, yet not extraordinary, human brain as a scaled-up primate brain and its associated cost. *Proceedings of the National Academy of Sciences*, 109(Supplement 1) (2012), pp. 10661–10668.
8. Worthy, G.A. and Hickie, J.P., Relative brain size in marine mammals. *The American Naturalist*, 128(4) (1986), pp. 445–459.
9. Herndon, J.G., Tigges, J., Anderson, D.C., Klumpp, S.A., and McClure, H.M., Brain weight throughout the life span of the chimpanzee. *Journal of Comparative Neurology*, 409(4) (1999), pp. 567–572.
10. Gabi, M., Collins, C.E., Wong, P., Torres, L.B., Kaas, J.H., and Herculano-Houzel, S., Cellular scaling rules for the brains of an extended number of primate species. *Brain, Behavior and Evolution*, 76(1) (2010), pp. 32–44.
11. Kazu, R.S., Maldonado, J., Mota, B., Manger, P.R., and Herculano-Houzel, S., Cellular scaling rules for the brain of Artiodactyla include a highly folded cortex with few neurons. *Frontiers in Neuroanatomy*, 8 (2014), p. 128.
12. Herculano-Houzel, S., Ribeiro, P., Campos, L., Da Silva, A.V., Torres, L.B., Catania, K.C., and Kaas, J.H., Updated neuronal scaling rules for the brains of Glires (rodents/lagomorphs). *Brain, Behavior and Evolution*, 78(4) (2011), pp. 302–314.
13. Cairó, O., External measures of cognition. *Frontiers in Human Neuroscience*, 5, (2011), p. 108.
14. Kinser, P.A., Brain and body size . . . and intelligence. Serendip Studio (2000). Available at: https://serendipstudio.org/bb/kinser/Int3.html [accessed 23 February 2021].

15. Seid, M.A., Castillo, A., and Wcislo, W.T., The allometry of brain miniaturization in ants. *Brain, Behavior and Evolution*, 77(1) (2011), pp. 5–13.
16. Pagel, M.D. and Harvey, P.H., Taxonomic differences in the scaling of brain on body weight among mammals. *Science*, 244(4912) (1989), pp. 1589–1593.
17. Fine, M.L., Horn, M.H., and Cox, B., *Acanthonus armatus*, a deep-sea teleost fish with a minute brain and large ears. *Proceedings of the Royal Society of London. Series B. Biological Sciences*, 230(1259) (1987), pp. 257–265.
18. Bahney, J. and von Bartheld, C.S., The cellular composition and glia-neuron ratio in the spinal cord of a human and a nonhuman primate: Comparison with other species and brain regions. *The Anatomical Record*, 301(4) (2018), pp. 697–710.
19. Huggenberger, S., The size and complexity of dolphin brains—a paradox? *Journal of the Marine Biological Association of the United Kingdom*, 88(6) (2008), p. 1103.
20. Armstrong, E. and Bergeron, R., Relative brain size and metabolism in birds. *Brain, Behavior and Evolution*, 26(3–4) (1985), pp. 141–153.
21. Herculano-Houzel, S., The remarkable, yet not extraordinary, human brain as a scaled-up primate brain and its associated cost. *Proceedings of the National Academy of Sciences*, 109(Supplement 1) (2012), pp. 10661–10668.
22. Olkowicz, S., Kocourek, M., Lučan, R.K., Porteš, M., Fitch, W.T., Herculano-Houzel, S., and Němec, P., Birds have primate-like numbers of neurons in the forebrain. *Proceedings of the National Academy of Sciences*, 113(26) (2016), pp. 7255–7260.
23. Kazu, R.S., Maldonado, J., Mota, B., Manger, P.R., and Herculano-Houzel, S., Cellular scaling rules for the brain of Artiodactyla include a highly folded cortex with few neurons. *Frontiers in Neuroanatomy*, 8 (2014), p. 128.
24. Gregory, T.R., Genome size and brain cell density in birds. *Canadian Journal of Zoology*, 96(4) (2018), pp. 379–382.
25. Jebb, D., Huang, Z., Pippel, M., Hughes, G.M., Lavrichenko, K., Devanna, P., Winkler, S., Jermiin, L.S., Skirmuntt, E.C., Katzourakis, A., and Burkitt-Gray, L., Six reference-quality genomes reveal evolution of bat adaptations. *Nature*, 583(7817) (2020), pp. 578–584.

26. Organ, C.L., Shedlock, A.M., Meade, A., Pagel, M., and Edwards, S.V., Origin of avian genome size and structure in non-avian dinosaurs. *Nature*, 446(7132) (2007), pp. 180–184.
27. Locke, J., *An Essay Concerning Human Understanding* (London: Kay & Troutman, 1847).
28. Hopping, M.E., Effect of light intensity during cane development on subsequent bud break and yield of 'Palomino' grape vines. *New Zealand Journal of Experimental Agriculture*, 5(3) (1977), pp. 287–290.
29. Ettinger-Epstein, P., Whalan, S., Battershill, C.N., and de Nys, R., A hierarchy of settlement cues influences larval behaviour in a coral reef sponge. *Marine Ecology Progress Series*, 365 (2008), pp. 103–113.
30. Orejas, C., Gori, A., Rad-Menéndez, C., Last, K.S., Davies, A.J., Beveridge, C.M., Sadd, D., Kiriakoulakis, K., Witte, U., and Roberts, J.M., The effect of flow speed and food size on the capture efficiency and feeding behaviour of the cold-water coral *Lophelia pertusa*. *Journal of Experimental Marine Biology and Ecology*, 481 (2016), pp. 34–40.
31. Mariscal, R.N. and Lenhoff, H.M., The chemical control of feeding behaviour in *Cyphastrea ocellina* and in some other Hawaiian corals. *Journal of Experimental Biology*, 49(3) (1968), pp. 689–699.
32. Hallock, M.B., Worobey, J., and Self, P.A., Behavioural development in chimpanzee (Pan troglodytes) and human newborns across the first month of life. *International Journal of Behavioral Development*, 12(4) (1989), pp. 527–540.
33. Bard, K.A., Platzman, K.A., Lester, B.M., and Suomi, S.J., Orientation to social and nonsocial stimuli in neonatal chimpanzees and humans. *Infant Behavior and Development*, 15(1) (1992), pp. 43–56.
34. Weisbecker, V. and Goswami, A., Brain size, life history, and metabolism at the marsupial/placental dichotomy. *Proceedings of the National Academy of Sciences*, 107(37) (2010), pp. 16216–16221.
35. Dyke, G.J. and Kaiser, G.W., Cracking a developmental constraint: egg size and bird evolution. *Records of the Australian Museum*, 62(1) (2010), pp. 207–216.
36. Nihei, Y. and Higuchi, H., When and where did crows learn to use automobiles as nutcrackers? *Tohoku Psychologica Folia*, 60 (2001), pp. 93–97.
37. Barash, D.P., Donovan, P., and Myrick, R., Clam dropping behavior of the glaucous-winged gull (*Larus glaucescens*). *The Wilson Bulletin*, 87 (1975), pp. 60–64.

38. Madrigal, A., Science can neither explain nor deny the awesomeness of this sledding crow. [online] *The Atlantic* (2009). Available at: https://www.theatlantic.com/technology/archive/2012/01/science-can-neither-explain-nor-deny-the-awesomeness-of-this-sledding-crow/251395/ [accessed 7 October 2020].
39. Sakura, O. and Matsuzawa, T., Flexibility of wild chimpanzee nut-cracking behavior using stone hammers and anvils: An experimental analysis. *Ethology*, 87(3–4) (1991), pp. 237–248.
40. Sousa, C., Biro, D., and Matsuzawa, T., Leaf-tool use for drinking water by wild chimpanzees (*Pan troglodytes*): Acquisition patterns and handedness. *Animal Cognition*, 12(1) (2009), pp. 115–125.
41. Sugiyama, Y., Drinking tools of wild chimpanzees at Bossou. *American Journal of Primatology*, 37(3) (1995), pp. 263–269.
42. Mann, J. and Patterson, E.M., Tool use by aquatic animals. *Philosophical Transactions of the Royal Society B: Biological Sciences*, 368(1630) (2013), p. 20120424.
43. Krützen, M., Mann, J., Heithaus, M.R., Connor, R.C., Bejder, L., and Sherwin, W.B., Cultural transmission of tool use in bottlenose dolphins. *Proceedings of the National Academy of Sciences*, 102(25) (2005), pp. 8939–8943.
44. Tebbich, S., Taborsky, M., Fessl, B., and Dvorak, M., The ecology of tool-use in the woodpecker finch (*Cactospiza pallida*). *Ecology Letters*, 5(5) (2002), pp. 656–664.
45. Barrett, G.A., Revell, D., Harding, L., Mills, I., Jorcin, A., and Stiefel, K.M., Tool use by four species of Indo-Pacific sea urchins. *Journal of Marine Science and Engineering*, 7(3) (2019), p. 69.
46. Hunt, G.R. and Gray, R.D., Direct observations of pandanus-tool manufacture and use by a New Caledonian Crow (*Corvus moneduloides*). *Animal Cognition*, 7(2) (2004), pp. 114–120.
47. Hunt, G.R., Tool use by the New Caledonian Crow *Corvus moneduloides* to obtain Cerambycidae from dead wood. *Emu-Austral Ornithology*, 100(2) (2000), pp. 109–114.
48. Martinho III, A., Burns, Z.T., von Bayern, A.M., and Kacelnik, A., Monocular tool control, eye dominance, and laterality in New Caledonian Crows. *Current Biology*, 24(24) (2014), pp. 2930–2934.
49. Chappell, J. and Kacelnik, A., Tool selectivity in a non-primate, the New Caledonian Crow (*Corvus moneduloides*). *Animal Cognition*, 5(2) (2002), pp. 71–78.

50. Weir, A.A., Chappell, J., and Kacelnik, A., Shaping of hooks in New Caledonian Crows. *Science*, 297(5583) (2002), pp. 981–984
51. Wimpenny, J.H., Weir, A.A., Clayton, L., Rutz, C., and Kacelnik, A., Cognitive processes associated with sequential tool use in New Caledonian Crows. *PLoS One*, 4(8) (2009), p. e6471.
52. von Bayern, A.M.P., Danel, S., Auersperg, A.M.I., Mioduszewska, B., and Kacelnik, A., Compound tool construction by New Caledonian Crows. *Scientific Reports*, 8(1) (2018), pp. 1–8.
53. Carryl, G.W., Fables for the Frivolous: (with Apologies to La Fontaine) (New York: Harper, 1899).
54. Hansen, W., The early tradition of the crow and the pitcher. *Journal of Folklore Research*, 56(2–3) (2019), pp. 27–44.
55. Bird, C.D. and Emery, N.J., Rooks use stones to raise the water level to reach a floating worm. *Current Biology*, 19(16) (2009), pp. 1410–1414.
56. Jelbert, S.A., Taylor, A.H., Cheke, L.G., Clayton, N.S., and Gray, R.D., Using the Aesop's fable paradigm to investigate causal understanding of water displacement by New Caledonian Crows. *PloS One*, 9(3) (2014), p. e92895.
57. Martinho III, A., Biro, D., Guilford, T., Gagliardo, A., and Kacelnik, A., Asymmetric visual input and route recapitulation in homing pigeons. *Proceedings of the Royal Society B: Biological Sciences*, 282(1816) (2015), p. 20151957.
58. Martinho III, A. and Kacelnik, A., Swapping mallards: Monocular imprints in ducklings are unavailable to the opposite eye. Animal Behaviour, 122 (2016), pp. 99–107.
59. Martinho III, A. and Kacelnik, A., Ducklings imprint on the relational concept of "same or different". *Science*, 353(6296) (2016), pp. 286–288.
60. Martinho-Truswell, A., McGregor, B., and Kacelnik, A., Ducklings imprint on chromatic heterogeneity. *Animal Cognition*, 22(5) (2019), pp. 769–775.
61. Taylor, J., Brown, G., De Miguel, C., Henneberg, M., and Rühli, F.J., MR imaging of brain morphology, vascularisation and encephalization in the koala. *Australian Mammalogy*, 28(2) (2006), pp. 243–247.

Chapter 4

1. Johnston, R.F., Siegel-Causey, D., and Johnson, S.G., European populations of the rock dove *Columba livia* and genotypic extinction. *American Midland Naturalist*, 120(2) (1988), pp. 1–10.
2. Murton, R.K. and Westwood, N.J., The foods of the rock dove and feral pigeon. *Bird Study*, 13(2) (1966), pp. 130–146.
3. Savard, J.P.L. and Falls, J.B., Influence of habitat structure on the nesting height of birds in urban areas. *Canadian Journal of Zoology*, 59(6) (1981), pp. 924–932.
4. Starck, J.M. and Ricklefs, R.E., Patterns of development: The altricial-precocial spectrum. *Oxford Ornithology Series*, 8 (1998), pp. 3–30.
5. Dial, K.P. and Jackson, B.E., When hatchlings outperform adults: Locomotor development in Australian brush turkeys (Alectura lathami, Galliformes). *Proceedings of the Royal Society B: Biological Sciences*, 278(1712) (2011), pp. 1610–1616.
6. Ricklefs, R.E., Avian postnatal development. *Avian Biology*, 7 (1983), pp. 1–83.
7. Shetty, S., Jacob, R.T., Shenoy, K.B., and Hegde, S.N., Patterns of breeding behaviour in the domestic pigeon. *Bird Behavior*, 9(1–2) (1990), pp. 14–19.
8. I should mention here that bird taxonomy is by no means a settled subject, and I am wary that the way I have simplified the relationship between bird groups here is undoubtedly going to leave some taxonomists upset with me. It is also very likely to become outdated, as the taxonomists rework and rearrange the various bird groups following more study.
9. Dunsworth, H. and Eccleston, L., The evolution of difficult childbirth and helpless hominin infants. *Annual Review of Anthropology*, 44 (2015), pp. 55–69.
10. Perry, J.S., The reproduction of the African elephant, *Loxodonta africana*. *Philosophical Transactions of the Royal Society of London. Series B, Biological Sciences*, 237 (1953), pp. 93–149.
11. Trevathan, W.R. and Rosenberg, K.R. eds., *Costly and Cute: Helpless Infants and Human Evolution* (Albuquerque, NM: University of New Mexico Press, 2016).
12. Bogin, B., The evolution of human childhood. *Bioscience*, 40(1) (1990), pp. 16–25.

13. Bogin, B., Evolutionary hypotheses for human childhood. *American Journal of Physical Anthropology: The Official Publication of the American Association of Physical Anthropologists*, 104(S25) (1997), pp. 63–89.
14. Rosenberg, K. and Trevathan, W., Bipedalism and human birth: The obstetrical dilemma revisited. *Evolutionary Anthropology: Issues, News, and Reviews*, 4(5) (1995), pp. 161–168.
15. Hirata, S., Fuwa, K., Sugama, K., Kusunoki, K., and Takeshita, H., Mechanism of birth in chimpanzees: Humans are not unique among primates. *Biology Letters*, 7(5) (2011), pp. 686–688.
16. Wong, K., Why humans give birth to helpless babies. *Scientific American* blog (2012), August 28.
17. Johnson, M.H., Functional brain development in humans. *Nature Reviews Neuroscience*, 2(7) (2001), pp. 475–483.
18. Duffield, D.A., Odell, D.K., McBain, J.F., and Andrews, B., Killer whale *(Orcinus orca)* reproduction at Sea World. *Zoo Biology*, 14(5) (1995), pp. 417–430.
19. Nerini, M.K., Braham, H.W., Marquette, W.M., and Rugh, D.J., Life history of the bowhead whale, *Balaena mysticetus* (Mammalia: Cetacea). *Journal of Zoology*, 204(4) (1984), pp. 443–468.
20. Nager, R.G., Monaghan, P., and Houston, D.C., The cost of egg production: Increased egg production reduces future fitness in gulls. *Journal of Avian Biology*, 32(2) (2001), pp. 159–166.
21. Visser, M.E. and Lessells, C.M., The costs of egg production and incubation in great tits *(Parus major)*. *Proceedings of the Royal Society of London. Series B: Biological Sciences*, 268(1473) (2001), pp. 1271–1277.
22. McLennan, J.A., Breeding of north Island brown kiwi, *Apteryx australis mantelli*, in Hawke's Bay, New Zealand. *New Zealand Journal of Ecology*, 11 (1988), pp. 89–97.
23. Warham, J., *The Petrels: Their Ecology and Breeding Systems* (London: A&C Black, 1990).
24. Bucher, T.L., Bartholomew, G.A., Trivelpiece, W.Z., and Volkman, N.J., Metabolism, growth, and activity in Adélie and emperor penguin embryos. *The Auk*, 103(3) (1986), pp. 485–493.
25. Franciscus, R.G., When did the modern human pattern of childbirth arise? New insights from an old Neandertal pelvis. *Proceedings of the National Academy of Sciences*, 106(23) (2009), pp. 9125–9126.

26. Linkenauger, S.A., Wong, H.Y., Geuss, M., Stefanucci, J.K., McCulloch, K.C., Bülthoff, H.H., Mohler, B.J., and Proffitt, D.R., The perceptual homunculus: The perception of the relative proportions of the human body. *Journal of Experimental Psychology: General*, 144(1) (2015), p. 103.
27. Almécija, S., Smaers, J.B., and Jungers, W.L., The evolution of human and ape hand proportions. *Nature Communications*, 6(1) (2015), pp. 1–11.
28. Metman, L.V., Bellevich, J.S., Jones, S.M., Barber, M.D., and Streletz, L.J., Topographic mapping of human motor cortex with transcranial magnetic stimulation: Homunculus revisited. *Brain Topography*, 6(1) (1993), pp. 13–19.
29. Padberg, J., Franca, J.G., Cooke, D.F., Soares, J.G., Rosa, M.G., Fiorani, M., Gattass, R., and Krubitzer, L., Parallel evolution of cortical areas involved in skilled hand use. *Journal of Neuroscience*, 27(38) (2007), pp. 10106–10115.
30. Hashimoto, T., Ueno, K., Ogawa, A., Asamizuya, T., Suzuki, C., Cheng, K., Tanaka, M., Taoka, M., Iwamura, Y., Suwa, G., and Iriki, A., Hand before foot? Cortical somatotopy suggests manual dexterity is primitive and evolved independently of bipedalism. *Philosophical Transactions of the Royal Society B: Biological Sciences*, 368(1630) (2013), p. 20120417.
31. Sutou, S., Hairless mutation: A driving force of humanization from a human–ape common ancestor by enforcing upright walking while holding a baby with both hands. *Genes to Cells*, 17(4) (2012), pp. 264–272.
32. McNutt, E.J., Zipfel, B. and DeSilva, J.M., The evolution of the human foot. *Evolutionary Anthropology: Issues, News, and Reviews*, 27(5) (2018), pp. 197–217.
33. Filler, A.G., Emergence and optimization of upright posture among hominiform hominoids and the evolutionary pathophysiology of back pain. *Neurosurgical Focus*, 23(1) (2007), pp. 1–6.
34. Weaver, T.D. and Hublin, J.J., Neandertal birth canal shape and the evolution of human childbirth. *Proceedings of the National Academy of Sciences*, 106(20) (2009), pp. 8151–8156.
35. Bogin, B., Evolutionary hypotheses for human childhood. *American Journal of Physical Anthropology: The Official Publication of the American Association of Physical Anthropologists*, 104(S25) (1997), pp. 63–89.
36. Hess, E.H., "Imprinting" in a natural laboratory. *Scientific American*, 227(2) (1972), pp. 24–31.

37. Shufeldt, R.W., Polygamy and other modes of mating among birds. *The American Naturalist, 41*(483) (1907), pp. 161–175.
38. Auld, J.R., Perrins, C.M., and Charmantier, A., Who wears the pants in a mute swan pair? Deciphering the effects of male and female age and identity on breeding success. *Journal of Animal Ecology, 82*(4) (2013), pp. 826–835.
39. Brugger, C. and Taborsky, M., Male incubation and its effect on reproductive success in the Black Swan, *Cygnus atratus. Ethology, 96*(2) (1994), pp. 138–146.
40. Black, J.M. ed., *Partnerships in Birds: The Study of Monogamy* (Oxford: Oxford University Press, 1996).
41. Chamberlain, M.J., Wightman, P.H., Cohen, B.S., and Collier, B.A., Gobbling activity of eastern wild turkeys relative to male movements and female nesting phenology in South Carolina. *Wildlife Society Bulletin, 42*(4) (2018), pp. 632–642.
42. Chakrabarti, S. and Jhala, Y.V., Battle of the sexes: A multi-male mating strategy helps lionesses win the gender war of fitness. *Behavioral Ecology, 30*(4) (2019), pp. 1050–1061.
43. Dunn, P.O. and Cockburn, A., Extrapair mate choice and honest signaling in cooperatively breeding superb fairy-wrens. *Evolution, 53*(3) (1999), pp. 938–946.
44. Double, M. and Cockburn, A., Pre-dawn infidelity: Females control extra-pair mating in superb fairy-wrens. *Proceedings of the Royal Society of London. Series B: Biological Sciences, 267*(1442) (2000), pp. 465–470.
45. Mulder, R.A. and Langmore, N.E., Dominant males punish helpers for temporary defection in Superb Fairy-wrens. *Animal Behaviour,* 45 (1993), pp. 830–833.
46. Lukas, D. and Clutton-Brock, T.H., The evolution of social monogamy in mammals. *Science, 341*(6145) (2013), pp. 526–530.
47. Stanford, C.B., The social behavior of chimpanzees and bonobos: Empirical evidence and shifting assumptions. *Current Anthropology,* 39(4) (1998), pp. 399–420.
48. Wrangham, R.W., The evolution of sexuality in chimpanzees and bonobos. *Human Nature,* 4(1) (1993), pp. 47–79.
49. Cockburn, A., Prevalence of different modes of parental care in birds. *Proceedings of the Royal Society B: Biological Sciences,* 273(1592) (2006), pp. 1375–1383.

50. Marlowe, F., Paternal investment and the human mating system. *Behavioural Processes*, 51(1–3) (2000), pp. 45–61.
51. Buss, D.M., Goetz, C., Duntley, J.D., Asao, K., and Conroy-Beam, D., The mate switching hypothesis. *Personality and Individual Differences*, 104 (2017), pp. 143–149.
52. Brown, L., Shumaker, R.W., and Downhower, J.F., Do primates experience sperm competition? *The American Naturalist*, 146(2) (1995), pp. 302–306.
53. Smith, N.W., Psychology and evolution of breasts. *Human Evolution*. 1 (1986), pp. 285–286.
54. Caro, T.M., Human breasts: Unsupported hypotheses reviewed. *Human Evolution*, 2(3) (1987), pp. 271–282.
55. Winfree, R., Cuckoos, cowbirds and the persistence of brood parasitism. *Trends in Ecology & Evolution*, 14(9) (1999), pp. 338–343.
56. Stoddard, M.C. and Stevens, M., Pattern mimicry of host eggs by the common cuckoo, as seen through a bird's eye. *Proceedings of the Royal Society B: Biological Sciences*, 277(1686) (2010), pp. 1387–1393.
57. Grim, T. and Honza, M., Does supernormal stimulus influence parental behaviour of the cuckoo's host? *Behavioral Ecology and Sociobiology*, 49(4) (2001), pp. 322–329.
58. Gangestad, S.W., Garver-Apgar, C.E., Simpson, J.A., and Cousins, A.J., Changes in women's mate preferences across the ovulatory cycle. *Journal of Personality and Social Psychology*, 92(1) (2007), p. 151.

Chapter 5

1. Zentall, T.R., Perspectives on observational learning in animals. *Journal of Comparative Psychology*, 126(2) (2012), p. 114.
2. Mather, J.A., Factors affecting the spatial distribution of natural populations of *Octopus joubini* Robson. *Animal Behaviour*, 30(4) (1982), pp. 1166–1170.
3. Wells, M.J. and Wells, J., Sexual displays and mating of *Octopus vulgaris* Cuvier and *O. cyanea* Gray and attempts to alter performance by manipulating the glandular condition of the animals. *Animal Behaviour*, 20(2) (1972), pp. 293–308.
4. Amodio, P. and Fiorito, G., Observational and other types of learning in Octopus. In *Handbook of Behavioral Neuroscience* (Vol. 22, pp. 293–302) (Oxford: Elsevier, 2013).

5. Fiorito, G. and Scotto, P., Observational learning in *Octopus vulgaris*. *Science*, 256(5056) (1992), pp. 545–547.
6. Boesch, C., Marchesi, P., Marchesi, N., Fruth, B., and Joulian, F., Is nut cracking in wild chimpanzees a cultural behaviour? *Journal of Human Evolution*, 26(4) (1994), pp. 325–338.
7. Matsuzawa, T., Biro, D., Humle, T., Inoue-Nakamura, N., Tonooka, R., and Yamakoshi, G., Emergence of culture in wild chimpanzees: Education by master-apprenticeship. In *Primate Origins of Human Cognition and Behavior* (pp. 557–574) (Tokyo: Springer, 2008).
8. Lefebvre, L., Brains, innovations, tools and cultural transmission in birds, non-human primates, and fossil hominins. *Frontiers in Human Neuroscience*, 7 (2013), p. 245.
9. Bluff, L.A., Kacelnik, A., and Rutz, C., Vocal culture in New Caledonian Crows *Corvus moneduloides*. *Biological Journal of the Linnean Society*, 101(4) (2010), pp. 767–776.
10. Hawkins, T.H., Opening of milk bottles by birds. *Nature*, 165(4194) (1950), pp. 435–436.
11. Lefebvre, L., The opening of milk bottles by birds: Evidence for accelerating learning rates, but against the wave-of-advance model of cultural transmission. *Behavioural Processes*, 34(1) (1995), pp. 43–53.
12. Sherry, D.F. and Galef, B.G., Cultural transmission without imitation: Milk bottle opening by birds. *Animal Behaviour*, 32(3) (1984), pp. 937–938.
13. Aplin, L.M., Sheldon, B.C., and Morand-Ferron, J., Milk bottles revisited: Social learning and individual variation in the blue tit, *Cyanistes caeruleus*. *Animal Behaviour*, 85(6) (2013), pp. 1225–1232.
14. Laurence, S. and Margolis, E., The poverty of the stimulus argument. *The British Journal for the Philosophy of Science*, 52(2) (2001), pp. 217–276.
15. Berwick, R.C., Pietroski, P., Yankama, B., and Chomsky, N., Poverty of the stimulus revisited. *Cognitive Science*, 35(7) (2011), pp. 1207–1242.
16. Chomsky, N., On cognitive structures and their development: A reply to Piaget. *Philosophy of Mind: Classical Problems/Contemporary Issues*, 751 (2006).
17. Pullum, G.K. and Scholz, B.C., Empirical assessment of stimulus poverty arguments. *The Linguistic Review*, 19(1–2) (2002), pp. 9–50.
18. Jarvis, E.D., Selection for and against vocal learning in birds and mammals. *Ornithological Science*, 5(1) (2006), pp. 5–14.

19. Esser, K.H., Audio-vocal learning in a non-human mammal: The lesser spear-nosed bat *Phyllostomus discolor*. *Neuroreport*, 5(14) (1994), pp. 1718–1720.
20. Ralls, K., Fiorelli, P., and Gish, S., Vocalizations and vocal mimicry in captive harbor seals, *Phoca vitulina*. *Canadian Journal of Zoology*, 63(5) (1985), pp. 1050–1056.
21. Sanvito, S., Galimberti, F., and Miller, E.H., Observational evidences of vocal learning in southern elephant seals: A longitudinal study. *Ethology*, 113(2) (2007), pp. 137–146.
22. Poole, J.H., Tyack, P.L., Stoeger-Horwath, A.S., and Watwood, S., Elephants are capable of vocal learning. *Nature*, 434(7032) (2005), pp. 455–456.
23. Foote, A.D., Griffin, R.M., Howitt, D., Larsson, L., Miller, P.J., and Rus Hoelzel, A., Killer whales are capable of vocal learning. *Biology Letters*, 2(4) (2006), pp. 509–512.
24. Lilly, J.C., Vocal mimicry in Tursiops: Ability to match numbers and durations of human vocal bursts. *Science*, 147(3655) (1965), pp. 300–301.
25. Bradbury, J.W. and Balsby, T.J., The functions of vocal learning in parrots. *Behavioral Ecology and Sociobiology*, 70(3) (2016), pp. 293–312.
26. Araya-Salas, M. and Wright, T., Open-ended song learning in a hummingbird. *Biology Letters*, 9(5) (2013), p. 20130625.
27. Beecher, M.D. and Brenowitz, E.A., Functional aspects of song learning in songbirds. *Trends in Ecology & Evolution*, 20(3) (2005), pp. 143–149.
28. Low, T., *Where Song Began: Australia's Birds and How They Changed the World* (New Haven, CT: Yale University Press, 2016).
29. Dalziell, A.H. and Magrath, R.D., Fooling the experts: accurate vocal mimicry in the song of the superb lyrebird, *Menura novaehollandiae*. *Animal Behaviour*, 83(6) (2012), pp. 1401–1410.
30. Putland, D.A., Nicholls, J.A., Noad, M.J., and Goldizen, A.W., Imitating the neighbours: Vocal dialect matching in a mimic–model system. *Biology Letters*, 2(3) (2006), pp. 367–370.
31. Robinson, F.N. and Curtis, H.S., The vocal displays of the lyrebirds (Menuridae). *Emu*, 96(4) (1996), pp. 258–275.
32. Sibley, C.G., The relationships of the lyrebirds. *Emu-Austral Ornithology*, 74(2) (1974), pp. 65–79.
33. Edwards, S.V. and Boles, W.E., Out of Gondwana: The origin of passerine birds. *Trends in Ecology & Evolution*, 17(8) (2002), pp. 347–349.

34. Atkinson, Q.D., Phonemic diversity supports a serial founder effect model of language expansion from Africa. *Science*, *332*(6027) (2011), pp. 346–349.
35. Moran, S. and McCloy, D., *PHOIBLE 2.0* (Jena: Max Planck Institute for the Science of Human History, 2019).
36. Naumann, C., The phoneme inventory of Taa (West !Xoon dialect). In *Lone Tree: Essays in Memory of Anthony Traill* (Cologne: Rüdiger Köppe, 2016).
37. Moran, S. and McCloy, D., *PHOIBLE 2.0* (Jena: Max Planck Institute for the Science of Human History, 2019).
38. Najar, N. and Benedict, L., The relationship between latitude, migration and the evolution of bird song complexity. *Ibis*, *161*(1) (2019), pp. 1–12.
39. Kaluthota, C., Brinkman, B.E., dos Santos, E.B., and Rendall, D., Transcontinental latitudinal variation in song performance and complexity in house wrens *(Troglodytes aedon)*. *Proceedings of the Royal Society B: Biological Sciences*, *283*(1824) (2016), p. 20152765.
40. Hurford, J.R., The evolution of the critical period for language acquisition. *Cognition*, *40*(3) (1991), pp. 159–201.
41. Newport, E.L., Bavelier, D., and Neville, H.J., Critical thinking about critical periods: Perspectives on a critical period for language acquisition. In *Language, Brain and Cognitive Development: Essays in Honor of Jacques Mehler* (pp. 481–502) (Cambridge, MA : MIT Press, 2001).
42. McWhorter, J., *What Language Is: And What It Isn't and What It Could Be* (Harmondsworth: Penguin, 2013).
43. Kibrik, A.E., Archi (Caucasian–Daghestanian). In *The Handbook of Morphology* (pp. 453–476) (Chichester: John Wiley & Sons, 2017).
44. Sizemore, M. and Perkel, D.J., Premotor synaptic plasticity limited to the critical period for song learning. *Proceedings of the National Academy of Sciences*, *108*(42) (2011), pp. 17492–17497.
45. Boughey, M.J. and Thompson, N.S., Song variety in the brown thrasher *(Toxostoma rufum)*. *Zeitschrift für Tierpsychologie*, *56*(1) (1981), pp. 47–58.
46. Price, P.H., Developmental determinants of structure in zebra finch song. *Journal of Comparative and Physiological Psychology*, *93*(2) (1979), p. 260.
47. Boettiger, C.A. and Doupe, A.J., Developmentally restricted synaptic plasticity in a songbird nucleus required for song learning. *Neuron*, *31*(5) (2001), pp. 809–818.

48. Nottebohm, F., Birdsong as a model in which to study brain processes related to learning. *The Condor, 86*(3) (1984), pp. 227–236.
49. Berwick, R.C., Okanoya, K., Beckers, G.J., and Bolhuis, J.J., Songs to syntax: The linguistics of birdsong. *Trends in Cognitive Sciences, 15*(3) (2011), pp. 113–121.
50. Slabbekoorn, H. and Smith, T.B., Bird song, ecology and speciation. *Philosophical Transactions of the Royal Society of London. Series B: Biological Sciences, 357*(1420) (2002), pp. 493–503.
51. Seddon, N. and Tobias, J.A., Song divergence at the edge of Amazonia: An empirical test of the peripatric speciation model. *Biological Journal of the Linnean Society, 90*(1) (2007), pp. 173–188.
52. Brainard, M.S. and Doupe, A.J., What songbirds teach us about learning. *Nature, 417*(6886) (2002), pp. 351–358.

Chapter 6

1. Groves, C., Primate taxonomy. In *The International Encyclopedia of Biological Anthropology* (pp. 1–6) (New York: Wiley, 2018).
2. Nater, A., Mattle-Greminger, M.P., Nurcahyo, A., Nowak, M.G., De Manuel, M., Desai, T., Groves, C., Pybus, M., Sonay, T.B., Roos, C., and Lameira, A.R., Morphometric, behavioral, and genomic evidence for a new orangutan species. *Current Biology*, 27(22) (2017), pp. 3487–3498.
3. Banks, W.E., d'Errico, F., Peterson, A.T., Kageyama, M., Sima, A., and Sánchez-Goñi, M.F., Neanderthal extinction by competitive exclusion. *PLoS One*, 3(12) (2008), p. e3972.
4. Ambrose, S.H., Volcanic winter in the Garden of Eden: The Toba supereruption and the late Pleistocene human population crash. *Volcanic Hazards and Disasters in Human Antiquity*, 345, (2000), p. 71.
5. Wynn, T., Piaget, stone tools and the evolution of human intelligence. *World Archaeology*, 17(1) (1985), pp. 32–43.
6. Gibson, K.R., Evolution of human intelligence: The roles of brain size and mental construction. *Brain, Behavior and Evolution*, 59(1–2) (2002), pp. 10–20.
7. Pinker, S., The cognitive niche: Coevolution of intelligence, sociality, and language. *Proceedings of the National Academy of Sciences*, 107(Supplement 2), (2010), pp. 8993–8999.

8. Sterelny, K., Social intelligence, human intelligence and niche construction. *Philosophical Transactions of the Royal Society B: Biological Sciences*, 362(1480), (2007), pp. 719–730.
9. Cosmides, L., Barrett, H.C., and Tooby, J., Adaptive specializations, social exchange, and the evolution of human intelligence. *Proceedings of the National Academy of Sciences*, 107(Supplement 2) (2010), pp. 9007–9014.
10. Berecz, B., Cyrille, M., Casselbrant, U., Oleksak, S., and Norholt, H., Carrying human infants: An evolutionary heritage. *Infant Behavior and Development*, 60 (2020), p. 101460.
11. Wrangham, R. and Conklin-Brittain, N., Cooking as a biological trait. *Comparative Biochemistry and Physiology Part A: Molecular & Integrative Physiology*, 136(1) (2003), pp. 35–46.
12. Tavares, E.S., Baker, A.J., Pereira, S.L., and Miyaki, C.Y., Phylogenetic relationships and historical biogeography of neotropical parrots (Psittaciformes: Psittacidae: Arini) inferred from mitochondrial and nuclear DNA sequences. *Systematic Biology*, 55(3) (2006), pp. 454–470.
13. Waterhouse, D.M., Parrots in a nutshell: The fossil record of Psittaciformes (Aves). *Historical Biology*, 18(2) (2006), pp. 227–238.
14. Waterhouse, D.M., Lindow, B.E., Zelenkov, N.V., and Dyke, G.J., Two new parrots (Psittaciformes) from the lower Eocene fur formation of Denmark. *Palaeontology*, 51(3) (2008), pp. 575–582
15. Christidis, L., Schodde, R., Shaw, D.D., and Maynes, S.F., Relationships among the Australo-Papuan parrots, lorikeets, and cockatoos (Aves: Psittaciformes): Protein evidence. *The Condor*, 93(2) (1991), pp. 302–317.
16. Joseph, L., Toon, A., Schirtzinger, E.E., Wright, T.F., and Schodde, R., A revised nomenclature and classification for family-group taxa of parrots (Psittaciformes). *Zootaxa*, 3205(2) (2012), pp. 26–40.
17. Young, A.M., Hobson, E.A., Lackey, L.B., and Wright, T.F., Survival on the ark: life-history trends in captive parrots. *Animal Conservation*, 15(1) (2012), pp. 28–43.
18. Wirthlin, M., Lima, N.C., Guedes, R.L.M., Soares, A.E., Almeida, L.G.P., Cavaleiro, N.P., de Morais, G.L., Chaves, A.V., Howard, J.T., de Melo Teixeira, M., and Schneider, P.N., Parrot genomes and the evolution of heightened longevity and cognition. *Current Biology*, 28(24) (2018), pp. 4001–4008.
19. Spoon, T.R., Parrot reproductive behavior, or who associates, who mates, and who cares. *Manual of Parrot Behavior*, 63 (2006), p. 77.

20. Martin, R.O., Long-term monogamy in a long-lived parrot: mating system and life-history evolution in the yellow-shouldered Amazon parrot *Amazona barbadensis*. (Doctoral dissertation, The University of Sheffield, 2009).
21. Spoon, T.R., Reproductive success, parenting, and fidelity in a socially monogamous parrot (cockatiels, *Nymphicus hollandicus*): The influence of social relationships between mates. (University of California, UCAL: X65187, 2003).
22. Bradbury, J.W. and Balsby, T.J., The functions of vocal learning in parrots. *Behavioral Ecology and Sociobiology*, 70(3) (2016), pp. 293–312.
23. Price, H., Birds of a feather talk together, Australian Geographic (2011) [online]. Available at: https://www.australiangeographic.com.au/news/2011/09/birds-of-a-feather-talk-together/ [accessed 15 October 2020].
24. Pepperberg, I.M., *The Alex Studies: Cognitive and Communicative Abilities of Grey Parrots* (Cambridge, MA: Harvard University Press, 2009).
25. Pepperberg, I.M., Numerical competence in an African gray parrot *(Psittacus erithacus)*. *Journal of Comparative Psychology*, 108(1) (1994), p. 36.
26. Pepperberg, I.M. and Gordon, J.D., Number comprehension by a grey parrot *(Psittacus erithacus)*, including a zero-like concept. *Journal of Comparative Psychology*, 119(2) (2005), p. 197–209.
27. Pepperberg, I.M., Social influences on the acquisition of human-based codes in parrots and nonhuman primates. In *Social Influences on Vocal Development* (pp. 157–177) (Cambridge: Cambridge University Press, 1997).
28. Pepperberg, I.M., *The Alex Studies: Cognitive and Communicative Abilities of Grey Parrots* (Cambridge, MA: Harvard University Press, 2009).
29. Pepperberg, I.M., Vocal learning in Grey parrots: A brief review of perception, production, and cross-species comparisons. *Brain and Language*, 115(1) (2010), pp. 81–91.
30. Kubiak, M., Feather plucking in parrots. *In Practice*, 37(2) (2015), pp. 87–95.
31. Jayson, S.L., Williams, D.L., and Wood, J.L., Prevalence and risk factors of feather plucking in African grey parrots (*Psittacus erithacus erithacus* and *Psittacus erithacus timneh*) and cockatoos (*Cacatua* spp.). *Journal of Exotic Pet Medicine*, 23(3) (2014), pp. 250–257.

32. Auersperg, A.M., Kacelnik, A., and von Bayern, A.M., Explorative learning and functional inferences on a five-step means-means-end problem in Goffin's cockatoos *(Cacatua goffini)*. *PloS One, 8*(7) (2013), p. e68979.
33. O'Hara, M., Mioduszewska, B., Mundry, R., Haryoko, T., Rachmatika, R., Prawiradilaga, D.M., Huber, L. and Auersperg, A.M., 2021. Wild Goffin's cockatoos flexibly manufacture and use tool sets. *Current Biology*.
34. Auersperg, A.M., Szabo, B., Von Bayern, A.M., and Kacelnik, A., Spontaneous innovation in tool manufacture and use in a Goffin's cockatoo. *Current Biology*, 22(21) (2012), pp. R903–R904.
35. Auersperg, A.M., von Bayern, A.M., Weber, S., Szabadvari, A., Bugnyar, T., and Kacelnik, A., Social transmission of tool use and tool manufacture in Goffin cockatoos (Cacatua goffini). *Proceedings of the Royal Society B: Biological Sciences, 281*(1793) (2014), p. 20140972.
36. Taylor, J., Chick flick: Cockatoo gives anti-nesting spikes the bird in viral video. [online] The Guardian (5 July 2019). Available at: https://www.theguardian.com/environment/2019/jul/05/chick-flick-cockatoo-gives-anti-nesting-spikes-the-bird-in-viral-video [accessed 15 October 2020].
37. NSW Environment, Energy and Science, How do I stop a cockatoo from attacking my property? (2018) [online]. Available at: https://www.environment.nsw.gov.au/questions/cockatoo-attack-property [accessed 15 October 2020].
38. Patel, A.D., Iversen, J.R., Bregman, M.R., and Schulz, I., Experimental evidence for synchronization to a musical beat in a nonhuman animal. *Current Biology, 19*(10) (2009), pp. 827–830.
39. Keehn, R.J.J., Iversen, J.R., Schulz, I., and Patel, A.D., Spontaneity and diversity of movement to music are not uniquely human. *Current Biology, 29*(13) (2019), pp. R621–R622.
40. Fitch, W.T., Biology of music: Another one bites the dust. *Current Biology, 19*(10) (2009), pp. R403–R404.
41. Schachner, A., Brady, T.F., Pepperberg, I.M., and Hauser, M.D., Spontaneous motor entrainment to music in multiple vocal mimicking species. *Current Biology, 19*(10) (2009), pp. 831–836.

42. White, N.E., Phillips, M.J., Gilbert, M.T.P., Alfaro-Núñez, A., Willerslev, E., Mawson, P.R., Spencer, P.B., and Bunce, M., The evolutionary history of cockatoos (Aves: Psittaciformes: Cacatuidae). *Molecular Phylogenetics and Evolution*, 59(3) (2011), pp. 615–622.
43. Murphy, S., Legge, S., and Heinsohn, R., The breeding biology of palm cockatoos (*Probosciger aterrimus*): A case of a slow life history. *Journal of Zoology*, 261(4) (2003), pp. 327–339.
44. Zdenek C.N., Heinsohn R., and Langmore N.E. Vocal individuality, but not stability, in wild palm cockatoos (*Probosciger aterrimus*). *Bioacoustics*, 27(1) (2018), pp. 27–42.
45. Zdenek, C.N., Who's who of palm cockatoos: Evaluating non-invasive techniques for identification of individual palm cockatoos (*Probosciger aterrimus*). (ANU, 2012).
46. Zdenek, C.N., Heinsohn, R., and Langmore, N.E., Vocal complexity in the palm cockatoo (*Probosciger aterrimus*). *Bioacoustics*, 24(3) (2015), pp. 253–267.
47. Heinsohn, R., Zdenek, C.N., Cunningham, R.B., Endler, J.A., and Langmore, N.E., Tool-assisted rhythmic drumming in palm cockatoos shares key elements of human instrumental music. *Science Advances*, 3(6) (2017), p. e1602399.
48. Dufour, V., Poulin, N., Curé, C., and Sterck, E.H., Chimpanzee drumming: A spontaneous performance with characteristics of human musical drumming. *Scientific Reports*, 5 (2015), p. 11320.
49. Clarkson, C., Jacobs, Z., Marwick, B., Fullagar, R., Wallis, L., Smith, M., Roberts, R.G., Hayes, E., Lowe, K., Carah, X., and Florin, S.A., Human occupation of northern Australia by 65,000 years ago. *Nature*, 547(7663) (2017), pp. 306–310.
50. Roberts, R.G., Jones, R., Spooner, N.A., Head, M.J., Murray, A.S., and Smith, M.A., The human colonisation of Australia: Optical dates of 53,000 and 60,000 years bracket human arrival at Deaf Adder Gorge, Northern Territory. *Quaternary Science Reviews*, 13(5–7) (1994), pp. 575–583.

FURTHER READING

Ackerman, J., 2021. *The Bird Way: A New Look at How Birds Talk, Work, Play, Parent, and Think.* London: Penguin.

De Waal, F., 2016. *Are We Smart Enough to Know How Smart Animals Are?* New York: WW Norton & Company.

Godfrey-Smith, P., 2016. *Other Minds: The Octopus and the Evolution of Intelligent Life.* London: HarperCollins.

Low, T., 2016. *Where Song Began: Australia's Birds and How They Changed the World.* New Haven, CT: Yale University Press.

McWhorter, J., 2013. *What Language Is: And What It Isn't and What It Could Be.* Harmondsworth: Penguin.

Pepperberg, I.M., 2008. *Alex & Me.* New York: HarperCollins.

Strycker, N.K., 2014. *The Thing with Feathers: The Surprising Lives of Birds and What They Reveal About Being Human.* New York: Riverhead Books.

LIST OF FIGURE CREDITS

Figure 4 – Nick Hobgood
Figure 6 – H. Raab
Figure 7 – © 2014 Grossi et al
Figure 9a – Photo by Elianne Dipp from Pexels
Figure 9b – Matthew T Rader/Unsplash
Figure 10 – Copyright © 2001-2021 Ellen Muller
Figure 11 – BBC Natural History
Figure 12 – Ole Jorgen Liodden / Nature Picture Library / Science Photo Library
Figure 13 – Behavioural Ecology Research Group Oxford
Figure 14 – Dr. Sarah Jelbert
Figure 15 – Antone Martinho-Truswell
Figure 16 – Peter Pevy
Figure 17 – Kseniia Glazkova / Alamy Stock Photo
Figure 18 – Imagenist/Shutterstock
Figure 19 – Blickwinkel / Alamy Stock Photo
Figure 20 – Storkalex/iStock
Figure 21 – Alice Auersperg
Figure 22 – Emma Martinho-Truswell
Plate 1 – Nimesh Madhavan/Flickr
Plate 2 – Peter Pevy
Plate 3 – Tracielouise/iStock
Plate 4 – James St. John, CC BY 2.0 <https://creativecommons.org/licenses/ by/2.0>, via Wikimedia Commons
Plate 5 – C. N. Zdenek

INDEX